29-50

QMC Library
23 1007281 0

DATE DUE FOR RETURN

21. JUN 1991
21. JUN 1991
26. JUN 1992
27 NOV 92
18. JAN 93
18. JUN 1993
20 SEP 93 Fm
26. NOV 93
17. JAN 94
31.

24 JAN 1995
1 - OCT 1999
28 FEB 2000
4 JUL 2000

WITHDRAWN
FROM STOCK
QMUL LIBRARY

The Evolution of Relativity

Christopher Ray

Adam Hilger, Bristol and Philadelphia

© IOP Publishing Ltd 1987

All rights reserved. No part of this publication may be reproduced, stored in a retrieval system or transmitted in any form or by any means, electronic, mechnical, photocopying, recording or otherwise, without the prior permission of the publisher.

British Library Cataloguing in Publication Data

Ray, Christopher
The evolution of relativity.
1. Relativity (Physics)——History
I. Title
530.1′1′09 QC173.52

ISBN 0-85274-423-4

Library of Congress Cataloging in Publication Data

Ray, Christopher.
The evolution of relativity.
Bibliography: p.
Includes index.
1. Relativity (Physics)——History. I. Title.
QC173.52.R39 1987 530.1′1′09 87-21137

ISBN 0-85274-423-4

Consultant Editor: **Professor A J Meadows**,
Loughborough University

Published under the Adam Hilger imprint by IOP Publishing Ltd
Techno House, Redcliffe Way, Bristol BS1 6NX, England
242 Cherry Street, Philadelphia, PA 19106, USA

Typeset by KEYTEC, Bridport, Dorset
Printed in Great Britain by J W Arrowsmith Ltd, Bristol

Contents

Introduction

John Archibald Wheeler, the renowned American physicist, once summed up Einstein's thought on relativity with these two simple yet powerful phrases: 'Matter tells space how to curve. And space tells matter how to move.' Einstein would have admired the elegance and economy of Wheeler's summary. For Einstein believed that simplicity is the key to the intelligibility of the physical world. He wrote, in 1933, 'Our experience up to date justifies us in feeling sure that in nature is actualized the ideal of mathematical simplicity.' For many, Einstein's General Theory of Relativity (GTR) embodies that ideal magnificently. The General Theory presents a view of gravitation in which the complexities of motion are deciphered by reference to *local* inertial frames. Even Newton had been suspicious of our tendency to rely on such global frames of reference as the 'fixed stars'. But he had replaced this material frame with another global frame, that of space itself. Einstein's genius was revealed in his realisation that local reference frames provide us with a new understanding of gravity and in his discovery that the fabric of space and time could have a dynamical role.

Relativity has stimulated the minds of many great physicists and has attracted the attentions of mathematicians—pure as well as applied; it has provided the raw material for many important historical studies; and it always challenged and sometimes confounded those with philosophical predispositions. But what is relativity theory? The usual response to such a question is to examine a popular text-book which represents the standard view of the subject: the Einstein field equations are seen as the focal point of a general theory with special relativity and Newtonian gravity being treated as special cases. This is certainly a satisfactory answer for the student who wishes to acquire the basic techniques which might enable him or her to resolve standard problems and puzzles. However, a deeper understanding yields far greater rewards. For the study of relativity

leads us on into cosmology, the physics of black holes, and quantum physics—into the heart of theoretical physics.

In this book I shall examine the fundamental simplicities of relativity; but I shall try to capture the wider significance of the theory. This investigation will help us to assess current accounts of the structure of science and to form a more precise view of physical theories and their development. Ernst Mach, Karl Popper and Thomas Kuhn have all had a dramatic effect, not just on how we understand the structures and development of science, but also on the attitudes to empirical and theoretical problems which are adopted by scientists. Mach demands a positivistic bias towards the observational and macroscopic side of physics; Popper asks us to proceed by formulating bold conjectures as physical hypotheses and then concentrate on rigorous attempts to refute these hypotheses—the method of falsificationism; and Kuhn warns us that science is not a simple accumulation of truths through successive theories, but that the real business of science is the resolution of problems and puzzles within the context of what he calls 'normal science'. Hence, those who explore the history and philosophy of science can produce prescriptions for scientific practice and not merely descriptions of actual behaviour and structures. Although such prescriptions and descriptions are sometimes impressive, they often incorporate a misleading view of scientific theories. Theories are frequently portrayed as relatively static structures which remain essentially the same throughout their lifetime. But a close look at relativity and its development will reveal a far from static organism. Relativity is an appropriate context for enquiry because it has high standing as a paradigm for 'classical' theoretical physics. Classical relativity is seen as a deterministic theory in the tradition of Newtonian mechanics and does not embrace automatically the revolutionary ideas of quantum physics.

My analysis begins with Newton's classical account of space and time and his characterisation of their absolute natures. Newton's ideas of space and time were attacked in the second half of the nineteenth century by Ernst Mach. This powerful and influential challenge led to the advocacy, initially by Einstein and then by others, of Mach's principle which repudiates the concept of absolute space–time. Because this principle is an important element in the development of relativity, I devote much of the first three chapters to its physical and philosophical rationale. Mach's principle is usually regarded as a questionable metaphysical fiat nowadays. But I suggest that we should view it as a methodological principle of simplicity. Ernst Mach's reputation has suffered much in recent years. However, I shall show that his thought provides a sound basis for our view of

science given his belief that simplicity is the 'fundamental conception of the nature of science' (Mach 1960: xxiii). Of course, I do not argue that Mach's principle has been vindicated by GTR but rather that it has played a crucial part in the evolution of relativity.

Chapters 2 and 3 deal with the classical exposition of GTR and the debate between absolutists and relationists which arises primarily from attempts to promote Machian ideas in the context of relativity. I examine the problems facing those, like Sciama, who are champions for Mach's principle. From the debates and disputes of classical relativity I move to the problems of singularities and the global structure of GTR space–times. Here we see one of the uneasy but effective marriages which physicists so often arrange in order to understand and explain the consequences of their theories. If we are to describe the behaviour of objects such as black holes in GTR space–times then the classical apparatus of relativity may be insufficient. The work of Stephen Hawking and others seems to require that quantum physics should enter the context of relativity theory. Despite attempts to maximise classical predictability—notably by Roger Penrose with his cosmic censorship hypothesis—quantum indeterminacy now seems to be an integral feature of space–time physics.

My review of classical relativity and its interactions with quantum physics leads me to present relativity as a dynamic, evolving theoretical context in which the debate and dissension that emerge in my analysis of the theory are central rather than peripheral features. The ideas of relativity are bound together in a network of mainstream and supporting variations. The coherence and fruitfulness of this network is assisted by principles and constraints which arise from considerations of simplicity. The search for simplicity has a fundamental influence not only on the content of a theory but also on the growth and development of the theory. Theories are not static structures—they evolve and adapt. But this evolution is by no means uncontrolled. The history of relativity from its birth in the early years of this century to its present amalgamation with quantum physics reveals both the fact of evolution and the primary means of control—the search for simplicity.

The ideas presented in this book emerged during research first in the Department of History and Philosophy of Science, Cambridge, and then at Balliol College, Oxford. I am indebted for the invaluable advice and assistance given to me during that research by Professor Mary Hesse, Dr Richard Healy and Dr Michael Fawcett in Cambridge, and by Dr Bill Newton-Smith and the late John Mackie in Oxford. Since then my arguments have benefited much from the criticism of many colleagues, students and friends. In particular,

Professor Chris Clarke of Southampton has helped me to clarify a number of important problems. Throughout Carol Ray has proved to be an exemplary critic.

This book is dedicated to Margaret Ray whose constant encouragement provided the foundation for my work. But I must also acknowledge the inspiration given to me by the writings of Ernst Mach and Stephen Hawking who both, with their powerful perspectives on the physical world, helped me to a better understanding of the role and nature of physics.

Christopher Ray
Oxford University

The long unmeasured pulse of time moves everything.
There is nothing hidden that it cannot bring to light,
Nothing once known that may not become unknown.
Nothing is impossible.

Sophocles, *Electra*
(transl. E F Watling)

Ernst Mach (1839–1916). Reproduced by permission of AIP Niels Bohr Library, *Physics Today* collection.

Chapter 1

Ernst Mach and the Search for Simplicity

Introduction

Ernst Mach's reputation has suffered in recent years. Philosophers have attacked his positivistic beliefs and his antagonism towards theoretical speculation. Historians of science have either played down Mach's influence on Einstein or ignored him altogether when discussing the origins of relativity, and many physicists have rejected his materialistic dynamics on the grounds that Mach's views are given little support by relativistic physics. All this may be a natural reaction to uncritical declarations that Einstein had vindicated Mach's approach to space, time and dynamics; see Russell (1927) and Reichenbach (1942, 1958). Einstein himself had become sceptical about Mach's positivistic beliefs whilst completing his work on the General Theory, and when he wrote his 'Autobiographical Notes', Einstein failed to endorse any complete enthusiasm for Machian ideas:

> . . . all physicists of the last century saw in classical mechanics a firm and final foundation for all physics . . . Even Maxwell and Hertz . . . adhered throughout to mechanics as the secured basis of physics. It was Ernst Mach who, in his Science of Mechanics, shook this dogmatic faith; this book exercised a profound influence upon me in this regard while I was a student. I see Mach's greatness in his incorruptible scepticism and independence; in my younger years, however, Mach's epistemological position also influenced me very greatly, a position which today appears to me to be essentially untenable. (1969: 21)

In order to assess the extent of Mach's influence on Einstein and to evaluate Einstein's verdict on Mach, we must begin with Mach's own views.

Mach's thought concerning the nature of space, time and dynamics comes to the forefront in his analysis of mechanics in *The Science of Mechanics* (1960). It also emerges in the early treatise on *Conservation of Energy* (1911), and receives attention in Mach's *Popular Scientific Lectures* (1943) and the collection of essays published under the title *Knowledge and Error* (1976). The highly positivistic and operationalistic philosophy of science which permeates these works is blended with Mach's own line in phenomenalism, which is presented with full force in *The Analysis of Sensations* (1959). Whilst I accept that Mach's philosophy of science is unpalatable, I believe this to be so only if we are asked to swallow the entire gamut of his ideas. I shall try to show that there are important features of Mach's work on dynamics which reveal much concerning the structure of theories and the practice of scientists; and that these features are central both to an understanding of Machian relationism and to Einstein's mature thought on space, time and gravitation.

If we are to build a complete picture of Machian relationism, then we must attend to the following aspects of Mach's thought:

(i) Mach's consideration of Newton's 'rotating vessel' or 'bucket' thought experiment;
(ii) his examination of Newton's 'two globes' thought experiment;
(iii) his critique of Neumann's thought on inertia;
(iv) his remarks concerning Corollary V of Newton's *Principia* (1729); and
(v) his views on the economy of science.

Most writers on the General Theory of Relativity (GTR) and Mach concentrate their attention upon the two experiments, (i) and (ii)—the former receiving the lion's share; see, for example, d'Abro (1927), Reichenbach (1958), Nagel (1961), Swinburne (1968), Jammer (1969), Lacey (1970), Earman (1970), Sklar (1974), Gardner (1977) and Zaret (1979). Of these authors, the most faithful to Newton's text is Jammer. One can only wonder why so many authors take for granted Newton's allegiance to absolute space in the Scholium. It is true, of course, that he begins by considering the nature of absolute space. Yet, when he comes to consider motion, Newton seems reluctant to make the straightforward claim that absolute rotation, for example, entails the existence of absolute space. However, Newton does not draw this inference. Others not only draw the inference, they also maintain that Newton does this. This is clearly an error. It is perhaps less of an offence to claim, as I shall do, that given Newton's argument the only reasonable interpretation of his account of the experiments (i) and (ii) is that he believed that absolute rotation is rotation relative to absolute space, even though he did not actually say so outright.

Few writers spend much, if any, time, in the context of Mach and GTR, on the ideas contained within (iii), (iv) and (v). A notable exception is Howard Stein in his 'Some philosophical prehistory of general relativity' (1977). Stein's brief but provocative study of Mach† confronts us with the assertion that Mach's discussion of space, time and motion reveals an acute loss of critical control. In making a survey of (i) to (v) I shall maintain, contrary to Stein, that Mach's thought on dynamics is largely coherent and consistent. The central characteristic of this thought is Mach's belief that economy is of fundamental importance in science.

1.1 Newton's thought experiments: the rotating vessel and the two globes

Before looking at Mach's critique of Newton I shall first set down a fairly detailed description of the two experiments, together with an account of the general argument in which they are employed. This will help us to highlight the strengths and weaknesses of Newton's views on space, time and motion as they appear in the Scholium following the definitions of the *Principia*.

1.1(a) The 'rotating vessel' or 'bucket' experiment

Newton believed that his rotating vessel or bucket experiment demonstrates that two kinds of motion obtain: absolute (or true) and relative (or apparent). The distinguishing feature is the presence of inertial forces which can be found only in bodies which are absolutely rotating. The device Newton used to illustrate this distinction was

> a vessel, hung by a long cord . . . (and) filled with water (1729: 15)

I shall not reproduce here the exact experiment as described by Newton, for it lacks the full force possessed by the following adaptation. I shall reconstruct the experiment in modern, not Newtonian, terms. The only important difference lies in the final stage which we shall examine: Newton did not seem to think that he needed this to make his point clear and so describes only the first three stages presented here.

Figure 1.1 depicts the four stages of the 'bucket' experiment, which we can consider to be taking place in a laboratory which is fixed relative to the distant stars, these providing at least an approximate inertial frame of reference.

†The study appears in an article covering Leibniz, Newton, Huygens, Kant, Helmholtz and Riemann, as well as Mach.

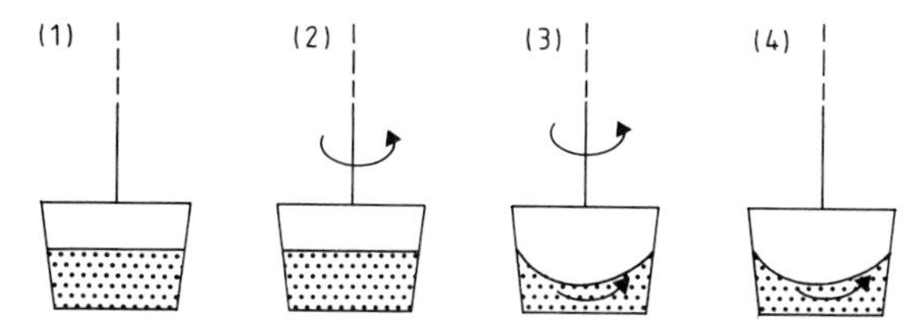

Figure 1.1 Newton's 'bucket' experiment.

(1) A bucket containing water is suspended with its axis of symmetry vertical; the bucket and water are both stationary with respect to the laboratory; the surface of the water is observed to be flat.

(2) The bucket is then set in rotation around the axis and relative to the laboratory; at this stage the water remains flat and at rest with respect to the laboratory.

(3) Gradually, the water creeps up the sides of the bucket, as it begins to spin with the bucket; maximum concavity of the water is reached when the bucket and water are at rest with respect to each other and spinning with respect to the laboratory.

(4) The bucket is then stopped suddenly but the water continues to swirl around rotating with respect to both bucket and laboratory, the surface of the water remaining maximally concave before receding slowly down the sides of the bucket.

Newton held that the curvature of the water in the third stage provides an observable and measurable consequence of inertial forces in operation. The reason for the concavity of the water is the centrifugal forces which 'push' the water away from the axis of rotation. Newton also claimed that the fact that the water experiences inertial forces is a clear sign that the water is in *absolute* and not merely relative motion. Newton's thought can be made clearer by attending to our 'final' stage of the experiment. We can see that at stage (4) as well as at stage (2) there is a maximum in relative rotation. If the inertial forces were due to merely relative motion between bucket and water one might expect the same effects at both these stages. But at stage (2) the water is flat, and curved at stage (4). The distinction which we are offered is this: the water at stage (2) is flat because the water, being in merely relative rotation to the bucket, experiences no inertial forces; the water at stage (4) although in relative rotation to the bucket is curved because it is in 'absolute rotation'. So, we may be persuaded to distinguish between the two kinds of motion because there are two distinct phenomena in this experiment: water with no inertial forces acting upon it, and water with inertial forces. Of course,

stage (3) also gives us a phenomenon of the latter type. Consequently, Newton may have felt this sufficed to distinguish absolute from merely relative rotation. Indeed, we can compare stage (3) with stage (1): the relative rotation here is identical in both cases, i.e. zero, yet only at (3) are there inertial forces in operation. Newton concluded that inertial forces do not arise out of merely relative rotation. Hence, Newton held that we are entitled to say that the water at stage (3) is in absolute rotation.

So, what is the effect of adding stage (4)? By doing so we show that in *all* cases of relative rotation, from zero to the maxima observed at stages (2) and (4), the same phenomena obtain. Although the comparison between stages (1) and (3) allows us to maintain that absolute rotation has real effects, it does not permit the claim that only absolute rotation has such effects. We need to rule out the possibility of any relative rotation giving rise to inertial forces. The comparison between stages (2) and (4) allows us to do this. Stage (4) enables us to generalise Newton's argument. We should also note that the relative rotations at stages (2) and (4) are in opposite directions, and we must further assume that this has no effect on the thought experiment (see Jammer (1969: 107) who makes this point). This assumption seems safe given that a second bucket, set rotating in the opposite direction and then stopped at the same time as the first bucket, could be expected to display the same behaviour.

The fourth stage of the thought experiment is therefore an important element in the argument for absolute space via absolute rotations. The primary purpose of the thought experiment is to establish the link between inertial forces and absolute rotations. Stage (4) makes this link clear-cut and prevents us from positing a further connection between inertial forces and some relative rotations. Although Newton did not make use of stage (4) in his argument, some commentators attribute it to him. Reichenbach (1958) seems to be the source of both the error and the extension of the experiment. Nagel (1961) and Horwich (1978) are two who follow Reichenbach's erroneous path. Although Jammer (1969) notes that Newton did not consider the final stage, he fails to say why its addition is important.

1.1(b) The 'two globes' experiment

Newton asks us to imagine two globes joined by a cord. When the globes are rotating about a common centre we will observe a tension in the connecting cord (see figure 1.2(*a*)); when there is no rotation there will be no tension (figure 1.2(*b*)).

We may begin by noting a number of riders to the above description. First, a gravitational force will be present, attracting the two globes together; this may be neglected for the purpose of the

experiment. Secondly, Mary Hesse has pointed out (Hesse 1961) that tension in the cord could be a sign of translational acceleration. I would add that there would also be tension if the system were to be experiencing both translational acceleration and rotation. But here we shall limit the discussion to rotation, as Newton does in his analysis of the two globes experiment.

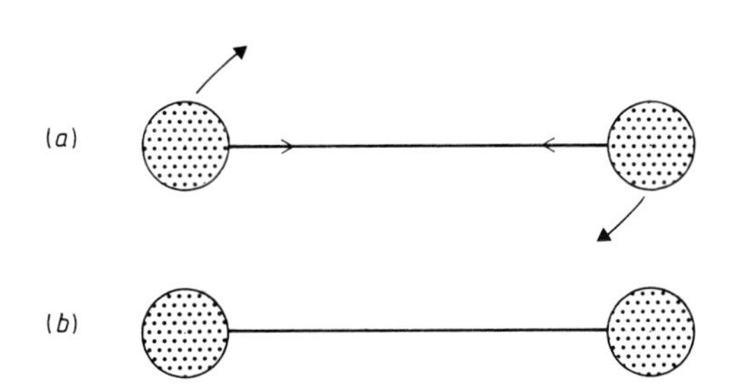

Figure 1.2 Newton's 'two globes' experiment.

Newton tells us that the tension as shown in figure 1.2(a) demonstrates the operation of inertial forces which are acting radially outwards from the centre of rotation. We can link rotation with these forces because in figure 1.2(b) there is no tension and no inertial force operating: the presence of the forces in figure 1.2(a) shows that the globes are rotating. But relative to what? Certainly not relative to each other: 1.2(a) and 1.2(b) are identical in this respect. We might be tempted to say: relative to the distant stars. If we fix our attention on these stars and on the globes, neglecting the cord, Newton (1729: 18) tells us that:

> We could not indeed determine from the relative translation of the globes among those bodies, whether the motion did belong to the globes or to the bodies. But, if we observed the cord, and found that its tension was that very tension which the motion of the globes requir'd, we might conclude the motion to be in the globes, and the bodies to be at rest.

Moreover, we might find the tension to be a sign of rotation:

> ev'n in an immense vacuum, where there was nothing external or sensible with which the globes could be compar'd.

Thus, Newton's argument proceeds as follows.

(1) In the two globe system there are two possible states of affairs: either there is tension in the connecting cord or there is not.

(2) Our general experimental findings in dynamics lead us to link such occurrences of tension with rotation.

(3) Therefore, if there is tension in the cord, the system is rotating.

(4) This rotation is not relative to any material frame of reference, for the same results as found in (2) can be expected even in an otherwise empty universe.

(5) Therefore, the rotation of the system is absolute.

There are two features of the argument which should be noted. First, the conditional form of step (4) which tells us that the results of the experiment would be no different even if there were no matter in the universe. Secondly, there is no further step alleging that absolute rotation entails the existence of absolute space. A number of commentators, for example Zaret (1979), present accounts of the experiment which erroneously maintain that Newton actually *says* that absolute space is the conclusion to which we must be led. As Cajori points out in Newton (1934), this is not the case. So, what allows us to infer absolute space from absolute rotation? In the bare descriptions of *both*† of Newton's experiments there is no reference at all to absolute space. Nor is any clear reason given for us to link absolute rotation with absolute space. Why, then, is it natural for most authors to see absolute space as the upshot of Newton's experiments? This can only be due to the influence of his earlier remarks in the Scholium, when he is 'defining' absolute and relative space, time and motion. There is a marked tendency for writers on this subject to concentrate on the issues raised by *either* these definitions *or* the bucket experiment *or* the two globes experiment. There are one or two exceptions to this unfortunate rule. Earman (1970) presents a brief but thorough account of the experiments and relates it to the earlier definitions. Sklar (1974) goes as far as to suggest that the conclusion of absolute space is the result of taking the two experiments together. I shall follow their example, merely adding that the Scholium possesses the structure of a *single* argument for absolute space, time and motion. This point is not made by Earman or Sklar, but seems evident when the Scholium is considered as a whole.

1.2 Newton's argument for absolute space

Now that we are armed with a clearer picture of the two experiments, we may set down the following argument, which follows Newton's reasoning in the Scholium step by step.

†Amongst others, Nagel (1961) believes that Newton's bucket experiment by itself is enough to establish absolute space.

(1) Material bodies move with respect to each other.

(2) Such motions result in changes in the relative position of bodies.

(3) *Definition*: absolute motion is motion from one absolute place to another.

(4) The absolute places taken together constitute a space which is fixed through time.

(5) *Definition*: relative motion is motion with respect to other bodies; a body in relative motion need not be in absolute motion.

(6) It is tempting to use material reference frames as though they were at absolute rest: such a frame is that of the distant 'fixed' stars.

(7) If they were actually at absolute rest, they would enable us to distinguish between absolute and merely relative motions: for any body moving with respect to the 'fixed' stars would then be in absolute motion; any body moving with respect to some other body but not with respect to the stars would possess only relative motion with respect to that other body.

(8) We could never know whether the fixed stars were actually at absolute rest, they may be moving at some constant velocity; Newton's laws entail that there is no observable difference between systems of bodies at rest or with a uniform linear motion of their centres of gravity.

(9) Therefore, the fixed stars do not give us a reliable way of distinguishing between all absolute and relative motions.

(10) Unless a force is applied to a body it will not change its state of motion.

(11) By the bucket experiment, bodies in relative rotation sometimes experience inertial forces, and sometimes do not.

(12) Those bodies which do experience such forces therefore change their state of motion and their position in the fixed space.

(13) Those bodies which experience no such forces cannot be said to be rotating absolutely.

(14) However, if the fixed stars are absolutely at rest or in uniform motion with respect to the fixed space, then absolute rotations can be observed relative to the material frame of the stars.

(15) By the globes experiment, the phenomenon of inertial forces does not depend on there being any material bodies other than that system affected by the forces.†

(16) Therefore, inertial forces alone are the distinguishing mark of absolute rotation.

(17) Consequently, material frames such as that of the stars (which might *seem* to act as an aid to distinguishing absolute and merely relative rotations) are superfluous.

†We should note that the bucket experiment carried out in empty space would provide equal support for this stage of the argument.

(18) Since there are inertial forces, there must be absolute motions.

(19) Since absolute motions are by definition motions from one absolute place to another, these absolute places must also exist.

(20) Given that the set of absolute places forms a fixed space, absolute motions are motions in this space, which is absolute space.

Apart from the last three steps I have followed the steps of the argument as advanced by Newton in the Scholium. But each of these three is obviously a reference back to one or more of the earlier steps.

There are two important features of this argument, upon which its strength depends. First, the linking of force with bodies *really* in motion (steps (10)–(18)). Secondly, the two attacks on the adoption of the fixed stars as *the* natural frame of reference for all motions (steps (6)–(9) and (14)–(17)). We shall now see how these features are highlighted by the two experiments.

When we take the argument as a whole, the part played by the two experiments becomes clear. It is also difficult to see how we can draw many conclusions merely from a consideration of one or other of the experiments. The last three steps depend upon earlier steps and act as a conclusion to the entire argument. Whatever significance we might attach to the fact that Newton did not finish the Scholium by stating these conclusions explicitly, it is clear that nothing hinders their assertion. But we can only do this on the foundation of the argument in its entirety. This is not to say that the two experiments do not have an important function in Newton's argument. The first helps us to see how force links with real as opposed to merely relative motion. It purports to establish that where an inertial force operates, there *is* an absolute rotation. But it is clear that Newton understood that this was not sufficient to establish either absolute rotation or absolute space. For he immediately admits that there are tremendous difficulties facing us when we try to distinguish absolute from merely relative motions. Although the presence of inertial forces gives us an important clue to the existence of absolute motions, there is still the problem of the fixed stars. These too could be used to distinguish absolute and relative motions. Newton attacks this notion in two ways. First, he notes that we cannot tell whether a body is at rest 'absolutely'. According to the 'Galilean' relativity principle, which Newton incorporated in his mechanics: the laws of mechanics hold good for all inertially moving systems. Given this principle, we are unable to decide which of two bodies in relative *uniform* motion is 'actually' moving. So, if we are in relative uniform motion to the stars, we could be at rest or they in motion, or vice versa, or we could both be in motion. Newton admits that the stars do appear to provide us with an inertial frame of reference. And this does give us the possibility of referring non-inertial motions such as rotations to this

frame. Consequently, Newton needs to show that such a material frame of reference is not a substitute for an absolute frame. The two globes experiment when performed in empty space is the device which Newton employs. With it, Newton tries to show that although the fixed stars are an apparent inertial frame, they are superfluous, for the phenomenon of inertial forces does not seem to depend upon the presence of external material bodies. So, Newton attacks first the idea that the distant stars provide a *genuinely* fixed frame for all motions and, secondly, the idea that the stars can be used to distinguish non-inertial from inertial motions. When these attacks are completed, and Newton sees no defence, he is then free to proclaim the existence of absolute space.

1.3 Newton and absolute time

The argument with which Newton leads us to the concept of space as a self-subsistent arena for real motion derives in part from his opposition to Descartes' views on space. The celebrated Leibniz–Clarke correspondence often leads readers to suspect that Newton's thought stands against the work of Leibniz, a champion of the relationist cause.† But, as Howard Stein points out, the context of the Scholium together with Newton's discussion of Descartes' doctrines on space and motion show that:

> Descartes was Newton's main philosophical target in the Scholium . . . Descartes' physics and cosmology constituted the most influential view in the scientific world at the time. (see Stein (1967: 184–5))

Stein supports this reasonable claim with persuasive textual evidence. An additional piece of evidence is the lack of any argument in the Scholium for absolute time. Newton simply states that real time is 'absolute' and 'from its own nature flows, equably without regard to anything external' (1729: 9). But in order to establish the idea of absolute space, Newton goes far beyond dogmatic statements, providing us with a powerful blend of empirical and philosophical reasoning, as seen above.

Why does Newton neglect the case for absolute time? Descartes' physics is concerned with the nature of space and motion; he defends, for example, the belief that the Earth is at rest—a belief which is refuted by Newton's laws of motion which immediately follow the Scholium. But Descartes has little to say about time to which Newton

†See Alexander's (1956) edition of the correspondence; Clarke was a staunch defender of Newton against Leibniz's relationist attacks.

might take exception. Newton's avowed interest is to use the *Principia* to combat 'certain prejudices' about dynamics (1729: 9). It seems likely that these prejudices arise from Descartes' influential ideas. If the Scholium were meant to provide us with a contrast to Leibniz's relational philosophy, then Newton would surely dwell rather more than he does on the problem of absolute time. For Leibniz—as shown in his correspondence with Clarke—has much to say on this subject. Hence, the lack of an argument for absolute time seems to add weight to Stein's contention that Newton addresses the Scholium to Descartes rather than to Leibniz.

Newton would not have found any problem in extending his argument for absolute space to cover absolute time as well. So long as we can abstract from the perceived physical universe to a model universe devoid of any material other than that of the state of affairs under investigation, as Newton does in the discussion of absolute space, then we can produce a Newtonian argument for absolute time. Newton's two globes thought experiment demonstrates that absolute space is needed if we are to differentiate between two possible circumstances: when there is no tension in the cord which links the two globes, and when there is tension. Empirical investigations show us that tension is an invariable sign of rotation in situations similar to that of the thought experiment: the same physical laws obtain for all such situations. But in a universe devoid of material frame of reference, any rotation must be taking place with respect to space itself.

To establish absolute time, we should begin with the claim that without material change there is no time. This claim is essentially relationist. A relationist would always point to changes in the order of material things to support his idea of time. We cannot measure time against any absolute scale; we need clocks—whether natural or man-made. The two globes experiment cannot be understood from this relationist perspective. If there is tension in the cord, then there is, on a Newtonian view, change. For the rotational acceleration which gives rise to the tension depends on the continuous change of velocity of each globe. Although the size of the velocities is constant, their directions are always changing. But, from a global point of view, there is no *material* change taking place. The tension is a sign of rotation; the rotation is a sign of change. And this change is a change *in time*. Since there is neither material reference frame nor clock for us to use, the time involved cannot be relative. It must be absolute: the continuous change of direction of the globes is taking place in absolute time. Hence, the relationist claim that without material change there is no time, seems to be ill-conceived. This argument depends on the same assumption as that for absolute space: we can make physical

laws universal and maintain that they apply in all appropriate circumstances regardless of any particular initial and boundary conditions. We shall now survey Mach's response to Newton's argument for absolute space and note his concern that this assumption may be unjustified.

1.4 Mach's critique of Newton

In §1.2, I argued that the Scholium has the structure of a single argument designed to establish the existence of absolute space, and that the two experiments we have examined must be seen in this context. Mach's most important attack upon Newton appears in Chapter II of *The Science of Mechanics* (1960) in his critique 'Newton's views of time, space and motion'. This central offensive is well supported by various sorties elsewhere in that book and in others. Mach does not give a detailed interpretation of the Scholium, but reproduces a large portion of it word for word. Much that Mach says is directed against the way Newton employs the two experiments and this seems to be because he saw these as particularly weak points in Newton's argument. Mach believes Newton's conclusion, that absolute space exists, is absurd, for:

> No one is competent to predict things about absolute space and absolute motion; they are pure things of thought, pure mental constructs, that cannot be produced in experience. All our principles of mechanics are . . . experimental knowledge concerning the relative positions and motions of bodies. (1960: 280)

This short passage sets the tone for most of Mach's attack on absolute space. The crucial contention that Mach must support is that absolute space and absolute motions do not occur in our experiences. By this Mach means that they have no observational or experimental effects. We have seen that the Scholium provides a dual challenge to this thesis. First it links inertial forces with absolute motions. Secondly, the claim that there is a natural material frame of reference for any motion is undermined. Newton does see a clear observational effect of some absolute motions—the inertial force; and he believes that such effects would occur even if external material bodies were no longer present. So, if Mach is effectively to oppose the concept of absolute space as held by Newton, he must face this challenge. He does so in a number of ways, and we shall look first at his remarks concerning the two experiments.

Mach agrees that inertial forces do exist but he maintains that the bucket experiment does not prove what Newton desires. It will be useful to examine Mach's analysis of Newton's claims in some detail. Mach contends that the facts allow us to say only that inertial forces always occur in bodies which are accelerated or rotating relative to astronomical frames of reference; bodies at rest or moving uniformly with respect to these frames experience no such forces. Mach tells us that the dynamic relativity between the fixed stars and, for example, a rotating body such as the Earth could be given two interpretations given that the inertial forces are caused by *relative* motion. Figure 1.3 demonstrates what is involved.

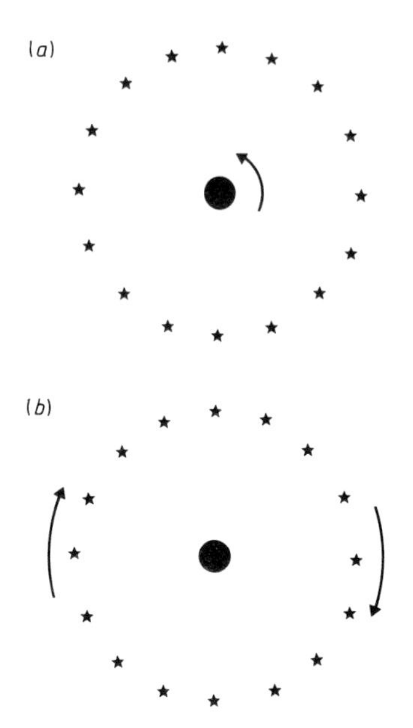

Figure 1.3 Relative rotation: the Earth and fixed stars.

Figure 1.3(*a*) shows the Earth rotating relative to the fixed stars; 1.3(*b*) shows us the stars rotating relative to a fixed Earth. Mach sets down what Newton says about the differences between 1.3(*a*) and 1.3(*b*) and then presents his own critique.

> If the earth is affected with an *absolute* rotation about its axis, centrifugal forces are set up in the earth: it assumes an oblate form, the acceleration of gravity is diminished at the equator, the plane of

> Foucault's pendulum rotates, and so on. All these phenomena disappear if the earth is at rest and the other heavenly bodies are affected with absolute motion round it such that the same *relative* rotation is produced. (1960: 284)

But Newton's account assumes the reality of absolute space and is not based on fact:

> . . . if we take our stand on the basis of facts, we shall find we have knowledge only of relative spaces and motions, (1960: 284)

The only circumstance we should be concerned with is the factual one, namely there is some dynamical state of affairs that obtains in the universe; we have knowledge of this state of affairs inasmuch as we can determine *observationally* or *experimentally* the relative motions between and forces amongst the bodies and systems of the universe. How we interpret this state of affairs is another matter. We might wish to say that the Earth is rotating or is fixed. Thus, we can see that Mach believes that the dynamical situation with regard to both possible states of affairs is identical. He believes that the same inertial forces would result since the relative rotation is identical in 1.3(*a*) and 1.3(*b*). Mach then applies this argument to Newton's bucket. He maintains that the *facts* concerning the second stage of the experiment,

> simply inform us that the relative rotation of the water with respect to the sides of the vessel produces *no* noticeable centrifugal forces. (1960: 284)

The surface only *appears* to be flat; we cannot know whether the relative rotation between the bucket and the water at this stage produces some *very slight* curvature of the water. If the stars affect the water, then the bucket itself might do so. Indeed, Mach suggests that if we had a tremendously large bucket with sides several miles thick, say, we might then observe the effects of the relative rotation easily. Two remarks that Mach makes support this broader view of dynamic relativity, which on Mach's admission goes beyond the facts available. First, on p.279 of *The Science of Mechanics* he says that '*all* masses and *all* velocities, and consequently *all* forces are relative', and on p.284 'The principles of mechanics can indeed, be so conceived that even for relative rotations centrifugal forces arise'. It is an empirical matter whether or not this latter conception is correct. I shall return to these ideas a little later.

The main point which Mach wants to make is this: we can explain the forces responsible for the curvature of the water by the 'relative rotation with respect to the earth and other celestial bodies' (1960:

284). That is, we have a perfectly respectable *material* frame of reference; we do not need to introduce, as the Scholium tries to do, extraordinary entities to explain the presence of inertial forces. The facts are: there is relative rotation; there are centrifugal forces. Mach tells us that the only reasonable conclusion is that one is linked with the other, that the latter is a dynamical effect of the former. Consequently, we may define the motion of Newton's bucket with respect to the rest of the universe; the fixed stars seem to provide us with an inertial frame of reference, so we should use them as such until we are proved to be in error. It is a mistake to maintain the existence of absolute motions or of absolute space. Thus, Mach contends:

> if we were to assert that we know more of moving objects than their . . . experimentally-given comportment with respect to the celestial bodies, we should render ourselves culpable of a falsity. When accordingly, we say that a body preserves unchanged its direction and velocity *in space*, our assertion is nothing more or less than an abbreviated reference to *the entire universe*. (1960: 286)

This gives expression to a limited view of dynamic relativity, namely that *the locally observed inertial forces are caused by acceleration or rotation relative to an inertial frame defined by the celestial bodies*. This view is 'limited' because Mach has nothing to say here about any effects upon the distant celestial bodies by the local non-inertial motions—he simply focuses upon the known local effects.

However, despite Mach's inclination to stick to the 'observable' facts, and, therefore, to relate inertial forces to the inertial frame of reference which the fixed stars apparently provide, Mach is sometimes much bolder. He makes some suggestive remarks about dynamic relativity which go beyond the facts, e.g. his conjecture that a suitably large bucket may produce measurable effects in the water. And these suggestions imply that Mach had in mind at least the *possibility* of a more complex theory of dynamics than the limited view above would allow by itself. Indeed we seem to be able to extend this limited viewpoint quite naturally. If the distant masses of the universe give rise to inertial forces in bodies moving non-inertially relative to them, then why should it not be the case that such a relative motion causes a like but tremendously small effect in the rest of the mass of the universe: hence, Mach's remark about a 'thick-sided' rotating bucket which might cause centrifugal forces in water at rest with respect to it. Just as the relative rotation of the bucket to the fixed stars (and the water) causes inertial forces within the bucket, so too this rotation might cause almost infinitesimal inertial forces in not only the water, but also the stars themselves! Mach believed (1943:

191; 1960: 243*f*) that Newton's crowning achievement was the law of the equality of action and reaction. The dynamical theory sketched here is certainly in accord with this law which Mach admired. One might expect the inertial effects of the stars on the bucket and water, and vice versa, to depend upon the respective masses and the distance between them. A possible relationship could be that proposed by Sciama (1961, 1969) in his 'simple' theory of gravitation and inertia. According to this theory the inertial force F between two masses m and M' distant r apart is given by

$$F = \frac{mm'}{r} \times a$$

where a is the relative acceleration between the masses. Hence, inertial forces are supposed to derive from the relative motions of masses: large masses such as the distant stars will induce inertial effects upon small bodies accelerating relative to them, and vice versa. Here we can see in sketch form one possible development of Mach's ideas. Of course, I do not pretend that Sciama's theory is an adequate explication of the phenomenon of inertia. I present it here merely to show the potential of this extension of Mach's ideas.

Of course, I might be accused of some liberality with Mach here, but it is not my intention to say exactly what Mach himself thought; but rather to show that his remarks on acceleration can be developed quite naturally to a more general theory of dynamic relativity. Moreover, this development is in accord with Mach's claim that *all* dynamical phenomena are relative. Indeed, Mach's reservations about the fixed stars as an inertial frame add further support to this development. Mach warns that we might not be able to hold the (limited) view of dynamic relativity 'if, for example, the requisite, relatively fixed bodies are wanting' (1960: 286). If it turned out that some astronomer had shown that the 'fixed' stars were not 'fixed' enough to provide an inertial frame of reference, what would Mach's reaction have been? I suggest that he would not have abandoned his philosophical commitment to relativity, but that he would have been more enthusiastic about his tentative ideas for a more general dynamics.

In table 1.1 I present three schemata; the first based upon Newton's interpretation of the bucket experiment, in the context of the Scholium as a whole, the second upon Mach's limited view of dynamic relativity, and the third upon the more general development that his views seem to suggest. These schemata show an analysis of the differences between stages (2) and (4) of the bucket experiment.

Stein (1977) accepts that Mach is committed to the 'limited' dynamical view which we have sketched in table 1.1. As we saw, according to this view there is no difference between the cases

Table 1.1

	Bucket		Water		Fixed Stars	
Schema 1 (Newton)						
Stage	(2)	(4)	(2)	(4)	(2)	(4)
Motion with relation to fixed stars	R	S	S	R	S	S
Inertial force experienced	M	—	—	M	—	—
Source of inertial force due to rotation relative to	Space	—	—	Space	—	—
Schema 2 (Mach)						
	Bucket		Water		Stars	
Stage	(2)	(4)	(2)	(4)	(2)	(4)
Motion	R	S	S	R	S	S
Inertial force	M	—	—	M	—	—
Source of inertial force	Stars	—	—	Stars	—	—
Schema 3 (generalised Mach)						
	Bucket		Water		Stars	
Stage	(2)	(4)	(2)	(4)	(2)	(4)
Motion	R	S	S	R	S	S
Inertial force	M	E	E	M	E	E
Source of inertial force	Stars (and water)	Water	Bucket	Stars (and bucket)	Bucket	Water

R, rotating; M, measurable; S, stationary or at rest; E, slight effect.

depicted by 1.3(*a*) and 1.3(*b*). Now, I have interpreted Mach's remarks concerning the 'thick-sided' bucket to provide us with an extension of the dynamics of these cases: the inertial force may be a two-way rather than just a one-way effect. But Stein believes that these tentative remarks are incompatible with his rather more definite views, as expressed in our Schema 2. Hence, Stein insists, we are being offered two incompatible physical theories by Mach.† The problem is this: Mach allows for the possibility that an experiment with a thick-sided bucket may have nil results. This might suggest that the two cases 1.3(*a*) and 1.3(*b*) are *not* equivalent. Inertial effects would then be observed on the Earth only if the Earth rotated relative to the fixed stars, and not if the Earth was fixed and the stars rotated around it. I agree that if Mach were committed to such a possibility, he would be guilty of an inconsistency. However, nothing Mach says

†Stein also detects a third theory which he sees in Mach's discussion of Corollary V of the *Principia*; I shall examine this 'discovery' later.

suggests that he inclines to a nil result for this experiment. And virtually everything we examined above supports the view that Mach would favour a positive result for the experiment. It is true that he asserts that 'no one is competent to say how the experiment would turn out' (1960: 284). But this did not seem to stop him from hoping that the result would not be negative. Mach's statement of allegiance to the relativity of motion and of force could hardly prevent him from doing so. Consequently, it is difficult to understand why Stein believes these remarks involve Mach in an inconsistency. Rather, as I have argued, they seem to complement his more definite ideas and provide us with a natural extension.

The foregoing analysis and schemata should help us to see how Mach's ideas and the extension of these ideas could operate in a practical application, and it is clear how these views differ from Newton's as regards the source of inertial forces. Whether we adopt the limited or the more general dynamical account, we must contradict Newton's story as presented in the Scholium. The inertial effects of the water are due to its *rotational* motion with respect to the stars.

I believe that Mach's attack on the bucket experiment does weaken Newton's argument for absolute space; for Mach is surely correct in arguing that we are given no reason to rule out material frames of reference in Newton's account of the bucket experiment. Earman (1970), however, argues that Mach's failure to extend and quantify his ideas into a full-blown theory of mechanics mitigates against any strength possessed by his criticism of the bucket experiment. It is clear to me that Mach is right in holding that the stars should not be dismissed as constituting the proper frame of reference for the bucket experiment. Of course, he does so not on the basis of any complex theory, but rather on the scantier foundation of a theory sketch. Nonetheless, I think that it is a strange view of scientific enquiry which prohibits us from challenging and perhaps overturning what seem to be faulty arguments merely because we do not have a *better* theory. Surely all we need are some theoretical ideas *worthy* of extension, correction and quantification. That Einstein was fired by Machian ideas is a clear indication of how worthy these ideas are.

We must remember, however, that the two globes experiment is a key step in Newton's argument; without it Mach's attack on the bucket experiment seems to win the day. It augments the bucket experiment's conclusion that inertial forces are linked with absolute rotations, and it seeks to preclude the fixed stars from being used as *the* frame of reference. The crucial feature in Newton's description of the two globes experiment, and that utilised in the overall argument for

absolute space, is the claim that the experiment can be performed in an otherwise empty universe. If this is so, then there is no *necessity* for material frames of reference; they become merely convenient aids in physical enquiry. Mach denies the feasibility of such experiments in both practice and principle. Stein maintains that Mach thereby exhibits some confusion. Stein tells us that:

> Mach says that we are 'not permitted to say' how things would be if the universe were differently arranged. But scientific theories ordinarily do permit all sorts of inferences about situations different from the actual one. (1977: 16)

However, we shall find that Mach is not so inflexible. Mach's denial of the feasibility of experiments like that of the two globes emerges most strongly in his consideration of Neumann's ideas regarding absolute space. Neumann adopts a similar approach to Newton's in the two globes experiment; he suggests that a body, rotating and experiencing centrifugal forces, would remain unaffected by the sudden disappearance of the rest of the universe. Mach retorts:

> when experimenting in thought, it is permissible to modify *unimportant* circumstances in order to bring out new features in a given case; but it is not to be antecedently assumed that the universe is without influence on the phenomenon here in question. (1960: 341)

Thus, Mach accepts the necessity for physicists to talk in terms of possibilities; but he imposes a limit upon their freedom to do so. Any assertion about a possible world *radically* different from the actual world—as in the case of talk about a universe empty apart from two globes and a string—is meaningless because it is observationally unverifiable. Who can tell what the nature of such a strange universe would be? It seems therefore that Stein is quite wrong in his claim above. Mach is only concerned with the prohibition of inferences from such radically different circumstances as might obtain in otherwise empty universes. Of course, we can complain that Mach is far from clear about the range of application of physical laws and about physical possibility. A standard way of dealing with this problem is in terms of possible worlds: a law may be said to be physically necessary when the law obtains in all worlds which differ from the actual world, if at all, only with respect to initial conditions; hence the law sets the boundaries of the physically possible (see, for example, Popper's *The Logic of Scientific Discovery* (1959: 420–41)). But such ways of trying to bring clarity to the problem of physical possibility will not help Mach. For we can easily imagine a possible world which is empty, apart from

two globes attached by a cord, and in which Newton's laws obtain! If we allow *any* changes to initial conditions, then we cannot prevent such possibilities. Mach wants a *restricted* view of physical possibility—but does not tell us where, nor how, to draw the line. This is a problem to which I shall return in Chapter 3, when we meet the problem of empty and near-empty space–times.†

Mach's own claim, that 'absolute' motion is actually motion relative to the entire universe, makes use of the observational fact that the distant 'fixed' stars do seem to provide us with something very close to an inertial frame of reference. For Newton, if the distant stars provide such a frame then this is contingent—a happy accident; accidental because space itself is the necessary inertial frame of reference (on the grounds of both thought experiments) and happy because the fixed stars help us in our measurements of astronomical movements. But, this contingency is transformed into a physical explanation of inertial forces by Mach.

However, the explanatory force of Mach's dynamic relativity is restricted by his failure to formulate the physical laws governing the effects of the large masses of the universe upon rotating and accelerating bodies 'and indeed governing relative motion in general'. There is also an additional problem. Mach's arguments, even if correct, do not refute the *possibility* of a *passive* 'container' which neither acts nor is acted upon (as opposed to Newton's active container which acts but is not acted upon). But this notion is physically empty, and Mach tells us that it is therefore physically meaningless. Since such a container would neither give us observational consequences not do any theoretical work, Mach's rejection seems justified.

Mach imputes reservations about the use of absolute space to Newton. Of course, Newton admits that

> the parts of space cannot be seen, or distinguished from one another by our senses; (1729: 12)

But, he goes on to maintain that

> in philosophical disquisitions, we ought to abstract from our senses, and consider things themselves, distinct from what are only sensible measures of them; (1729: 12)

And, as we have seen, the fixed stars for Newton are merely a sensible measure of inertial forces; the real measure is absolute space. Yet, Mach points out in later editions of *The Science of Mechanics* that

†I shall have more to say on Mach's views on physical possibility at that juncture.

Newton's Corollary V 'does not refer to absolute space'.† Mach contends that this frequently used corollary 'alone has scientific value' (1960: 340).

> In order to have a generally valid system of reference, Newton ventured the fifth corollary . . . he imagined a momentary terrestial system of co-ordinates, for which the law of inertia is valid, held fast in space without any rotation relative to the fixed stars . . . by this view Newton gave the *exact* meaning of his hypothetical extension of Galileo's law of inertia. We see that the reduction to absolute space was by no means necessary, for the system of reference is just as relatively determined as in every other case. In spite of his metaphysical liking for the absolute Newton was correctly led by the *tact of the natural investigator.* (1960: 285)‡

Stein maintains that Mach's support for Corollary V leads him into difficulties. We have already seen that Mach believes there to be no dynamical difference between the Earth rotating with respect to the fixed stars and the stars rotating around a fixed Earth. We have also tried to answer some of Stein's criticisms of Mach. But here, too, Stein holds that Mach is inconsistent. For, Stein tells us, Corollary V is a statement of Galilean relativity and makes no reference at all to the fixed stars, the 'cosmic masses' (1977: 18). Since this statement is quite compatible with a rotation of these distant masses about the Earth, but with no inertial effects being induced, Stein believes that Mach is committed to yet another physical theory. In this there is a difference between the two cases of rotation above. Whilst one or two of Mach's comments on Corollary V§might lead us to believe that he is attached to it as just a statement of Galilean relativity, the passage I have set down above makes it quite obvious that Mach sees Corollary V in the context of the fixed stars.|| Stein says Mach makes no reference to the cosmic masses, but Mach understands Newton to be referring the inertial frames of Corollary V to the fixed stars. In the passage above, Mach says this explicitly. Mach sees evidence for his interpretation of Newton, saying that Newton, in his discussion of

†Newton's Corollary V states: 'The motion of bodies included in a given space are the same among themselves, whether that space is at rest, or moves uniformly forwards in a right line without any circular motion'. (1729: 30)

‡Further evidence of Mach's enthusiasm for Corollary V appears in the Notes to the *Conservation of Energy* (1911: 95).

§ For instance in the Preface to the seventh edition of *The Science of Mechanics* (1960: xxviii).

||A passage in *Knowledge and Error* (1976: 345) also makes this clear.

planetary motions, refers them to the fixed stars. Mach seems to be quite justified: in the third part of the *Principia* Newton does refer the motion of first the solar system and then the Moon to the fixed stars (1934: 574, 580). Thus, Mach's commitment here is not just to Corollary V, but to this corollary *and* the fixed stars. Stein, however, might well say this is inconsistent. He maintains that Corollary V despite its lack of any reference to absolute space still relies implicitly on absolute motion. If this were true, then Mach's adherence to the fifth corollary would entail an inconsistent attachment to the absolute! However, if we remember the argument that Newton advances for absolute space, the mere fact that a body is moving inertially and is therefore suffering no force (see step (10)) does not imply anything about the absolute. Yet, the set of inertial frames have a privileged status for Newton—for it is only with respect to these that his laws are valid. But, we need to link inertial forces with absolute motion. Newton's restatement of the principle of Galiliean relativity is simply concerned with the status of the inertial motions. More information is required if we are to say that anything which suffers inertial forces is not moving inertially *and* is therefore in absolute motion. In attempting to establish this final conclusion Newton used the bucket and two globes experiment. But if we believe, as Mach does, that Newton's argument is imperfect, then we may conclude that material frames of reference might not be superfluous. They would consequently provide us with at least a sketch of an explanation for *why* the inertial frames have a privileged status. As Sciama points out

> absolute space does not really provide a physical explanation for the preferred role of inertial frames. Rather, it is simply a restatement of their existence. (1969: 14)

If we can show that inertial forces have their source in matter acting upon matter, as Mach believed, then we can explain why inertially moving bodies are force-free. This would also help to show why the inertial frames satisfy the principle of Galilean relativity. All we need is a theory to bind these ideas together.

This survey of Mach's critique of Newton has enabled us to see the key elements of Mach's thought on dynamics. Although Mach lacks anything like a full theory of inertia and gravitation, a number of conditions for such a theory are advanced by Mach. In summary, the essential points are as follows.

(1) Space has no independent 'absolute' existence.

(2) It is only physically meaningful to talk in terms of the material objects in the universe and their properties and interrelations.

(3) All motion is relative to at least one other object or system.

(4) The inertial forces which we observe may be explained by the fact that they arise only in objects accelerating or rotating with respect to fixed celestial or material frames of reference.

As a final point I add the substance of my suggested development of Mach's dynamical viewpoint: as set down in table 1.1, Schema 3.

(5) We should not exclude the possibility of inertial forces arising in any body or system accelerating or rotating relative to any other body or system: the detection of such forces may only be prevented by the insensitivity of the appropriate measuring instruments.†

The first three points here present Mach's epistemological challenge to absolute space, and the last two present an empirical challenge.

1.5 Mach, his phenomenalism and the economy of science

Mach's philosophy of science has two distinct aspects: phenomenalism and the commitment to economy, or simplicity, as a guiding rule in scientific thought. Although Mach's full-blown phenomenalism is now deemed to be philosophically unsound, simplicity provides us with the key to understanding the structure and evolution of scientific theories, as I shall argue in the final chapter. But simplicity is also an important element in Mach's rejection of absolute space and in Einstein's approach to gravitation.

Mach's phenomenalistic philosophy of science incorporates a good measure of operationalism. This seems to guarantee his resistance to the notion of absolute space: the only things that we may count as 'real' are the phenomena which we encounter in observation and experiment; our theories about the physical world should simply be shorthand accounts of these phenomena. 'Theoretical' explanations are eschewed. Thus, we are asked not to talk in terms of space itself—which Mach believes we do not, and cannot meet in sensory experience—but about the phenomena we do experience; and these, for Mach, are material objects and their properties and interrelationships.

Mach's approach has a good deal in common with the later philosophies of Bridgman (operationalism) and of Carnap (positivism).‡ Two features link Mach's ideas with these philosophies of

†In 1894 Friedlander tried to resolve this problem with fairly sensitive apparatus; his results were inconclusive.

‡See Bridgman (1936) and Carnap (1936) for classical expressions of these philosophies.

science. First, there is a common commitment to observation and experiment as *the* fundamental aspect of scientific discourse—observation statements are basic and their meanings, when analysed correctly, are unproblematic. Secondly, there is a common distrust of theoretical statements—so much so that only when such statements can be precisely defined in observational terms are they admissible in genuinely scientific discourse. Zahar (1981: 268) sums up these two elements elegantly with two 'empiricist' principles:

> (1) the observation level is both low and contingently infallible;

and

> (2) scientific hypotheses should in the last analysis be propositions *about* the phenomenal level only

According to the second principle, theoretical discourse is to be discouraged. This is because theoretical terms refer to unobservable entities. The adoption of Zahar's two empiricist principles involves a commitment to the now notorious 'theory/observation dichotomy', and anyone who adopts them must face the accusation that he or she fails to recognise that 'observation is theory-laden'. But we must be careful to distinguish between two issues here: (i) the division of scientific discourse into two parts—observational and theoretical; and (ii) the classification of entities and events as observable and unobservable. As van Fraassen (1980: 14) warns us, we should not commit category mistakes with talk of (un)observable terms or concepts or of theoretical entities; but we should separate questions about scientific language from questions about the entities referred to by that language. Amongst those who take pains to distinguish between these questions are Hesse (1974), Papineau (1979) and van Fraassen (1980). When asked what the thesis that observation is theory-laden means, we might therefore make two responses: our sense experience (of 'observable' entities) depends upon our theories (this is argued notably by Hanson (1958)); and the language we use to report sense experience (i.e. our 'observation' language) depends upon our theories (this is forcefully argued in Sellars (1963) and Feyerabend (1981)).

The strength of the attack on the 'theory/observation dichotomy' lies in the claim that our 'observation' language is infected by theory—however low-level that language might be. Even if, contrary to Hanson, we accept that it is possible to have a purely 'observational' experience—of a patch of red, for instance—the truth of any report of that experience is not guaranteed itself, since the association of the report with the experience may be at fault. And what underwrites

such an association? Not merely the experiences involved, but also the more general beliefs, views etc, which we hold, together with any tests we might devise to check that a correct association has been made. As Papineau points out:

> The way we associate words with experiences seems always to require assessment by reference to our general theories; (1979: 28)

And van Fraassen adds his support:

> All our language is thoroughly theory-infected. . . . The way we talk, and scientists talk, is guided by the pictures provided by previously accepted theories. This is true also, as Duhem already emphasized, of experimental reports; (1980: 14)

That is, even the simplest observation report presupposes a theoretical context. As Hesse says, this has drastic consequences for science:

> . . . no feature in the total landscape of functioning of a descriptive predicate is exempt from modification under pressure from its surroundings. That any empirical law may be abandoned in the face of counter examples is trite, but it becomes less trite when the functioning of every predicate is found to depend essentially on some laws or other and when it is also the case that any 'correct' situation of application—even that in terms of which the term was originally introduced—may become incorrect in order to preserve a system of laws and other applications; (1974: 16)

Now, if it is true—as it seems—that our observation language is theory-laden in the manner specified above, then our observation reports can no longer be seen as having a validity which is guaranteed simply by the sense experience about which we are reporting. We must look also to the theoretical context, i.e. to the system of predicates and laws of which our report is a part. If we are to have clarity and precision in our observation reports, the theoretical system must be precisely formulated. And if that system involves essential references to 'unobservable' entities like absolute space, then we may, in certain circumstances, come to doubt the validity of an observation report because it conflicts with a law which carries 'theoretical' reference to an 'unobservable' entity. This seems to be precisely what Newton wants us to do when he asks us to doubt the validity of the observation report that it is only in non-inertial motion with respect to the fixed stars that inertial forces may arise. For Newton's laws involve a commitment to absolute space which Newton readily admits does not 'come under the observation of our senses' (1729: 17). If we accept Newton's laws and also wish to accept the counterfactual assertion that they would hold even in the absence of the distant stars,

then we must challenge the report that inertial forces arise from relative motions alone. That is, a commitment to the laws and their general application brings us to doubt the validity of an observation report concerning relative motions. Of course, Mach answers Newton by demanding a new set of laws which involve no allegiance to unobservable entities. Indeed, if Mach could present us with a system of predicates and laws in which 'theoretical' terms played no part, his 'observation' statements would be theory-laden, but perhaps not disastrously so. For the observation statements would be theory-laden in the sense outlined above, namely they would depend on the system as a whole for their validity. But any observation report made would not depend on 'theoretical' terms, for these terms *ex hypothesi* play no part in the system. Hence, Mach might then argue that he can get along without such theoretical terms as 'absolute space'. Of course, he would not be able to contend that observation statements are unproblematic, especially if the theoretical context is complex. But how cogent would Mach's case be? Can science dispense with theoretical terms, like 'is a neutrino', 'is a field' or 'is a black hole', and make do with observational terms like 'is red', 'is cold' or 'is ten units long'? The phenomenalist programme demands that our descriptions of physical objects should be in terms of sensory experience, being set down in statements about the objects of that experience, namely sense data. And if a purported 'object' cannot be characterised in terms of sense data statements, then that object is 'theoretical' and must, in the final analysis, be excluded from scientific discourse. Mach refers to the objects of sensory experience as 'elements'; he says

> The world consists of colours, sounds, temperatures, pressures, spaces, times, and so forth, which we shall not call sensations, nor phenomena, because in either term an arbitrary, one-sided theory is embodied, but simply elements. The fixing of the flux of these elements, whether mediately or immediately, is the real object of physical research. (1943: 209)

Mach avoids the terms 'sensations' and 'phenomena' simply in order to stress his belief that experience of the world is not subjective, but objective, in the way that sensory experience is experience *of* the world and can be characterised in an objective manner.

The problems of phenomenalism are well known, and I shall merely set down here two or three of the difficulties which advocates of that programme must face.† First, the possibility of a 'sense data'

†For full details, see, for example, Nagel (1961: 121*f*), Hempel (1965: ch.5) and Papineau (1979: ch.1).

language is very remote indeed—no one has succeeded, or even come near to succeeding, in the construction of such a language. Secondly, whilst one might be willing to regard talk of elementary particles as suspicious until characterised in phenomenal terms (should it prove possible), it seems rather far-fetched to suspect talk of desks, pens and paper and other such everyday objects merely because this talk is not couched in sense data terms. Thirdly, if two people hold even slightly different general views concerning the physical world, there exists the possibility of a disagreement over the validity of an observation report and hence over the content of a sense datum statement; this follows from our acceptance that observational terms are theory-laden, even when the 'theory' held might be devoid of theoretical terms. These and other problems present a formidable barrier to any naive phenomenalist approach to science. Hence, one might be inclined to agree with those philosophers who, like Earman, maintain that

> the extreme forms of positivism and operationalism are not a suitable basis on which to found an adequate epistemology. (1970: 298)

Even so, I shall try to show that Mach's philosophy of science is not altogether barren.

In *The Evolution of Scientific Thought* d'Abro remarks that we must not understand 'phenomenalism' in any narrow sense if we are to find a place for it in relativity. We must therefore question the extreme to which Mach went, e.g. in denying

> the existence of atoms merely because they had never been observed by human eyes, regardless of whether it was useful to conceive of them for the purpose of co-ordinating empirical facts. This is not the attitude of science. But, if by phenomenalism we mean the desire to free our understanding of things from *unnecessary* metaphysical notions which are in no way demanded by experiment, then we are undoubtedly justified in claiming that not only the theory of relativity, but modern science itself, is essentially phenomenological. (1927: 431)

It is gratifying to note that d'Abro sees that even modern science has a metaphysical content; but the implication of this passage is that phenomenalism in its 'wider' sense is essentially that approach to science which demands that metaphysical notions must be kept to a minimum. Mach is roundly condemned for his excesses. His rigid conception of what counts as real encourages him to deny existence to those entities which are not part of our sensory experience. Material bodies are the 'stuff' of this 'direct' experience; but atoms and absolute space, highly complex theoretical constructions, are dismissed as physically empty notions.

The problem that I wish to resolve is this: what is the source of Mach's antagonism to absolute space? Earman and Stein maintain that it is solely Mach's brand of phenomenalism which is to blame, and they, as well as d'Abro, believe that Mach's attitude to the absolute is just as much a product of this philosophy as is his anti-atomism. Earman tells us that

> Mach is widely praised for his rejection of absolute space but is criticised for his rejection of atomism. But both of these rejections stem from exactly the same source; namely his highly positivistic and operationalistic philosophy. (1970: 298)

Their interpretation of Mach's anti-atomism is on the same lines as that presented by Blackmore in his biography *Ernst Mach* (1972): the roots of Mach's attitude to the atomic theories of the nineteenth century can be traced to his phenomenalistic philosophy. Yet this view is not free from difficulties.

Mach's philosophy of science was hardly well formed when he first objected to atomic theory. So, if we are to connect Mach's early anti-atomism with his phenomenalism, then it is with a very ill-defined philosophy. The final sentence of the *Conservation of Energy*, written in 1872, gives voice, nevertheless, to his early views:

> The object of natural science is the connexion of phenomena; but the theories are like dry leaves which fall away when they have long ceased to be the lungs of the tree of science. (1911: 74)

And Mach pronounces that physics is 'The discovery of the laws of the connexion of sensations (perceptions)' (1911: 91). However, Mach does not develop these ideas to any great extent until after the publication of the *The Science of Mechanics* (1960) in 1883. The main vehicles for this development were *The Analysis of Sensations* (1959) and *Knowledge and Error* (1976), both published in the early twentieth century. Only in these does Mach *openly* admit his philosophical motives. In the preface to *Knowledge and Error* Mach announces that the book is a study in 'scientific methodology' and 'the psychology of knowledge' (1976: xxxii); and, in *The Analysis of Sensations*, he sets out to examine 'the philosophical viewpoint of the average man' (1959: 37). Laudan contends that the fact that we had to wait for many years to see these mature works is good enough reason to dispense with the notion that it was Mach's phenomenalism which formed his adverse opinions on atomism. But the two quotations from Mach's earlier works above reveal that phenomenalism is an evident feature of his thought. That this must have had at least some influence upon Mach's approach to atomism is made clear by another passage in the

Conservation of Energy. Here Mach is referring to earlier writings on atomic theory:

> In the year 1862, I drew up a compendium of physics for medical men, in which, because I strove after a certain philosophical satisfaction, I carried out rigorously the mechanical atomic theory. This work first made me conscious of the insufficiency of this theory . . . I was busied, at the same time with psycho-physics, and so I became convinced that the intuition of space is bound up with the organisation of the senses, and, consequently, that we are not justified in ascribing spatial properties to things which are not perceived by the senses. In my lectures on psycho-physics I already stated clearly that we are not justified in thinking of atoms spatially. (1911: 87)

Despite this clear sign that ideas concerning sense experience did play a part in Mach's anti-atomism, I believe that Blackmore and his allies are wrong to put all their eggs into this basket. The raw form of phenomenalism which appears in his early works is merely a part of a general approach to science. Here I am persuaded by Hiebert that Mach's view of science incorporated the ideas of Mach the scientist, of Mach the historian and of Mach the philosopher (Hiebert 1970, 1976(a,b)). A characteristic assertion from Hiebert is this:

> In my opinion all of Mach's historico-critical writings support the thesis that the history of science was for him a tool to interpret and illuminate epistemological problems in the philosophy of science that he was perplexed about as a physicist. (1976(a): 376–7)

This view certainly seems to make sense. Mach's greatest contribution to physics was *The Science of Mechanics*. Not a text-book, nor a treatise offering a new synthesis of physical ideas, this book is a combination of physics, history and philosophy, with the emphasis on the first two elements; but, despite Mach's protestations that he was no philosopher,† the last element also plays an important part. With such a wide approach to science as is presented there, it would hardly be surprising to find this attitude reflected throughout his work. Indeed, Mach says of *The Science of Mechanics*:

> The present volume is not a treatise upon the application of the principles of mechanics. Its aim is to clear up ideas, expose the real significance of the matter, get rid of metaphysical obscurities. . . . The gist and kernal of mechanical ideas has in almost every case grown up in the investigation of very simple and special cases of mechanical processes; and the analysis of the history of the discussions concerning these cases must ever remain the method at once the most effective and

†See, for example, Mach (1976: xxxii).

> the most natural for laying this gist and kernal bare. Indeed, it is not too much to say that it is the only way in which a real comprehension of the general upshot of mechanics is to be attained. (1960: preface to first German edition, xxii–xxiii)

In addition, as Hiebert points out, in the opening pages of the *Conservation of Energy* Mach stresses the need for a historical approach to science. So I find it hard to dissuade myself from accepting Hiebert's conclusion. Yes, phenomenalism does play a large part in Mach's attitudes to science in general and atomism in particular; but his experience as an experimental physicist, and his critical writings as an historian of science must also have been involved. So, when we consider Mach's views on space, time and motion, we must see his rejection of absolute space in this context.

Mach's general approach to science is, however, tempered by one central feature, a dominant idea. Mach tells us that his 'fundamental conception of the nature of science' is the 'economy of thought' (1960: preface, xxiii). For Mach,

> The goal (of physical science) is the *simplest* and *most economical* abstract expression of the facts. (1943: 207)

This is the essence of his theory of the economy of thought. Although Mach maintains that the economy of thought first became an *important* element of his thinking in 1882/3† (with the publication of *The Science of Mechanics*, and his lecture 'On the economical nature of physical thinking'‡), he tells us that this conception was in his mind as early as 1861.§

Blackmore presents a comprehensive but misleading account of Mach's theory of economy. With eleven separate quotations Blackmore hopes to show us the scope and content of Mach's thought on this subject. He covers the following: economy of thought; economy of energy; economy of work and time; methodological economy; mathematical simplicity; abbreviation; abstraction; logical economy; ontological economy; economy in nature; and linguistic economy (1972: 173–4). However, Blackmore offers no analysis of the part economy plays in Mach's philosophy of science; nor does he investigate whether the above kinds of economy have any interrelationships. I shall do three things here:

(i) I shall show that Mach's economy of science need be classified

†See *The Analysis of Sensations* (1959: 49).

‡This essay appears in *Popular Scientific Lectures* (1943).

§ See *The Science of Mechanics* (1960: 591).

under three headings only: the economy of form, content and labour;

(ii) I shall examine the role of the theory of economy in Mach's view of science; and

(iii) I shall mention a number of difficulties that Mach's theory must face.

In the polemical work *Force and Matter* Mach's contemporary, Ludwig Büchner, maintained that: '*simplex veri signillum*'—'simplicity is the hall-mark of the truth' (1884: 14).† Mach would never have subscribed to this maxim. Although considerations of simplicity and economy are of the utmost importance for science and for scientists, the 'truth' is not so easily approached. For reality, nature, and therefore the truth about these, are complex matters. So, how can the scientist acquire knowledge? Mach tells us that economy of thought is the key to our understanding of nature. But he warns us in a Humean voice that the regularities that we might impose on nature for simplicity's sake are not to be found in nature herself, they are a construction of the mind.‡ To achieve this simplicity, we will assume that nature is uniform; for example, if we do not suppose that the heavens possess 'great relative stability' we could have no hope of forecasting the movements of the stars, for 'with great changes in celestial space, we should lose our co-ordinate system of space and time' (1943: 206). Yet we have no cause to believe that this assumption is unjustified. Mach's guiding rule, perhaps the closest to any concise statement of a 'principle' of economy, instructs us to regard

> science . . . as a minimal problem, consisting of the completest possible presentment of facts with the least possible expenditure of thought. (1960: 586)

And throughout Mach's thought on economy we can find much advice on how this goal can be attained.

First, we must recognise that the world of experience is far too complex for us to reproduce in full. The mind cannot handle such intricacies, so it employs its instruments: memory, abstraction and the symbolisation of language and mathematics.

> Experiences are analysed, or broken up, into simpler and more familiar experiences, and then symbolised at some sacrifice of precision. (1960: 578)

†Einstein was introduced to Büchner's *Force and Matter* in 1889 by Max Talmud; see Pais (1982: 520).

‡Mach acknowledges his debts to Berkeley and Hume in *The Analysis of Sensations* (1959: 367–8).

and,

> Memory is handier than experience, and often serves the same purpose. (1960: 577)

but,

> Because the mental power, the memory of the individual is limited, the material of experience must be arranged. (1911: 55)

Moreover

> We never reproduce the facts in full, but only that side of them which is important to us, moved to this directly or indirectly by a practical interest. Our reproductions are invariably abstractions. (1960: 578–9)

These devices help us to communicate effectively and build up a collective knowledge. Our main interest in using them is to maximise the information for a minimum cost in effort. With them we can economise on labour with a 'great saving of time and avoidance of fatigue' (1960: 584). We can then begin to order the facts of experience scientifically. But because our knowledge is limited by our inadequacies, our need to economise pushes us toward the construction of theories:

> We fill out the gaps in experience by the ideas that experience suggests. (1960: 588)

All these ideas suggest that without *the economy of labour* in science we will be unable to advance to an adequate comprehension of the world.

We also need to attend to the form in which our ideas are presented. By using the symbols of language and mathematics we can set down ideas concisely. Mach maintains that

> Mathematics is the method of replacing in the most comprehensive and economical manner possible, new numerical operations by old ones done already with known results. (1960: 583)

And, with mathematics we have realised

> The greatest perfection of mental economy. (1943: 195)

Mechanics, too, is highly developed economically because its facts are reduced to just a few elements: spaces, times and masses. Hence the combination of mechanics with mathematics has good reason to

be regarded as a paradigm of sciences, not in the sense that we should seek to reduce all sciences to mechanics, but rather that it offers a model in its *form* for other sciences. The equations of mechanics and other sciences help us to simplify complex relationships which obtain between phenomena. By removing such complexity we help to produce explanations of the physical world. Mach holds that the *economy of form* is essential for our understanding and explanation of the world, and aids the economy of labour:

> Language, with its helpmate, conceptual thought, by fixing the essential and rejecting the unessential, constructs its rigid pictures of the fluid world on the plan of a mosaic, at a sacrifice of exactness and fidelity but with a saving of tools and labor. (1943: 192)

The symbols of language and mathematics are the main instruments of both economy of form and economy of labour.

Mach also directs us to work towards an *economy of content*. Even the simplest theorems of mechanics represent the complexities of nature; and, therefore, we cannot expect to proceed in science without difficulty:

> scientific research is somewhat like unravelling complicated tangles of string, in which luck is almost as vital as skill and accurate observation. (1976: 10)

Now, strictly speaking, the fruits of such research will be *thoughts* about facts, about our direct experience of the world. As we saw above, these thoughts are best expressed in abstract and symbolic form. The content of thought therefore is not the facts themselves, but our simplified representations of these facts. Our theories should be built upon this foundation. But, Mach warns us,

> Not all the prevalent scientific theories originate so naturally and artlessly; (1960: 588)

We must beware the tendency to give way to practical myths and to accept artificial entities as real phenomena. We should avoid the dangers of metaphysical thinking. An example of such thought is the belief that atoms exist. In *The Analysis of Sensations* Mach depicts atoms as hypothetical entities which may have a limited purpose in science:

> [atoms] remain economical ways of symbolizing experience. But we have as much right to expect from them, as from the symbols of algebra, more than we have put into them, and certainly not more enlightenment than from our experience itself. (1959: 311)

Mach seems to imply that metaphysical entities like atoms and absolute space are *empty* symbols: they have no experiential content. There does seem to be an important difference in Mach's approach to atoms and to Newton's absolute space. Mach allows the former, if only for their instrumental value. But Mach is fastidious in his resistance to absolute space, denying that it has any place in science. The only possible reason for this appears to be that Mach sees atoms as artificial 'entities' which do have some scientific functions but absolute space which has no place in science at all is relegated to the realm of poetical mythology. But both atoms and absolute space seem to serve just as adequately as hypothetical entities which are mere instruments of science. Mach's lack of clarity here is matched by the general disorder of his thinking concerning theoretical entities. Atoms and absolute space are singled out for harsh treatment, but many nineteenth-century conceptions which appear to be the same kind of theoretical devices escape scot-free, for example oxygen or gravitational force.† Perhaps Mach would tell us that these latter two are admissible because they are grounded in experience, whereas the former two notions are not so based. And, certainly, the bulk of Mach's polemic against atoms and absolute space endeavours to show the tenuous links which these entities have with the phenomenal world. Nonetheless, all such devices are used for their economical features: they are distinguished into two groups—those with content based on fact, and those without such content. If we are to achieve economy of content then the clear implication of Mach's admonitions is that we should minimise the extent to which our theories need to incorporate artificial devices. This will help us to gain a clearer understanding of the world, since the language of our theory will be a more perfect representation of the facts. We shall also be able to manipulate the ideas contained in the theory more easily, since we will not be encumbered with an excess of content. Consequently, we shall facilitate the economics of labour and of form. However, we shall not be able to dispense with such entities until our understanding becomes complete. They will be needed to fill in the gaps, until the time

> when the closed circuit of physical and psychological facts shall lie complete before us. (1943: 212)

Mach's economy of thought therefore encompasses labour, form and content. Together they give us a unified idea: our scientific endeavours should be made with the minimum effort for the maximum return.

†Laudan makes a similar point to this (1976: 392).

Mach's theory of economy cuts much deeper than Occam's razor, which is concerned with content alone: superfluous entities and kinds of entities should be excluded from our ontology. Mach sees economy as a requirement which should be given wider dominion. Mach's theory is quite general and can apply to any branch of science or mathematics. We need not be concerned directly with phenomena: Mach's claim that mathematics is a model of economy makes this clear. Yet, the demands of economy cannot be divorced from those of phenomenalism in Mach's thought. When Mach required that the content of scientific theories be non-metaphysical, he is fulfilling the requirement that the content should be expressed in as economical a form as possible: we are cutting down on the kinds of entity to which our theories refer. The demand for phenomenalism, however, seems far narrower than that of economy. We are given no instructions about methods or about the tools of science which we should adopt. We are only offered the bald and unhelpful statement that we should stick to observation and experiment alone. If we must advance theories, then they must be based firmly on experience; and this experience belongs to the (unpalatable) realm of sensation. Mach's theory of economy, on the other hand, provides us with advice on many aspects of science and scientific research. Although Mach gives voice to his theory in phenomenal language, his ideas on economy seem to run deeper than those on phenomena, and Mach's theory of economy seem to provide the motivation for his phenomenalism. Although we might be persuaded to accept that phenomenalism plays a vital role in the economy of content, it could not do so in the economy of form and of labour.

We can see how phenomenalism is motivated and constrained by the theory of economy. When we look to matter as the basis of the phenomenal world we are doing so with interests of economy at heart; when we rule out absolute space from our ontology, we do so for reasons of economy; and when we assume that the distant stars are 'fixed' and the movements of celestial bodies are uniform, economy prevails. Of course, Mach's phenomenalism is also evident in all these circumstances. Yet the theory of economy is also concerned with non-phenomenal matters. Hence, it seems to be reasonable to take Mach's theory of economy as more fundamental than his phenomenalism. Indeed, Mach himself proclaims, economy is the *fundamental* conception of science. The incongruence of the respective domains of economy and phenomenalism certainly support Mach's proclamation; phenomenalism is concerned only with the content of our theories, but economy encompasses all aspects of theories and theorising.

Unfortunately, Mach's theory of economy suffers from one impor-

tant defect. Throughout there is a naive assumption that economy is a simple matter; that the most economical form of a theory is a straightforward idea. One example seems to be sufficient to dispel this presumption. There are several methods for evaluating the area under a curve; they are Simpson's rule, Euler's method and the trapezoidal method. Presumably Mach would ask us to choose that method which produced the most accurate answer. This is, of course, Simpson's rule. Mach would also want us to use a method which would minimise our labour: and this would be either the trapezoidal method or Simpson's rule, depending on the equations we are handling. Finally, Mach would like us to employ the simplest possible form for our equation: and this is provided by Euler's method. So, Mach's demands of economy for content, labour and form could direct us to three different methods! Mach fails to mention that there may be conflict between competing requirements of economy.

If Mach's ideas on economy are to provide a coherent basis for scientific investigations and theories, we must therefore add a rider to his theory. We must seek a balance when the various economies compete. There is no point in legislating in general just how this balance should be arrived at. It will depend upon the particular scientific (or mathematical) circumstances, and perhaps upon the capabilities of, and instrumental aids available to, the scientist.

Mach claims pre-eminence for himself as an advocate for the economy of science (1960: 592); nonetheless, his theory of economy not only suffers from the difficulty mentioned above, it is also remarkably woolly. If the aim of science is to produce comprehensive representations of the phenomenal world in as economic a manner as possible, then we will need to know when the goal of economy has been reached. Mach gives us no firm criteria which might serve this purpose. He merely offers us general advice and points to mathematics as a paradigm. But even with mathematics he is not precise enough to evade the problem of conflicting economies, nor does he attempt to separate subjective ideas of economy from objective ideas. This latter distinction, which Hesse makes (1974: 223), is important if we are to be in a position to give rigorous expression to our criteria of economy. Subjective economies, made for pragmatic or psychological reasons, will not lend themselves easily to such rigour. Mach's failure to make this distinction certainly contributes to the vagueness of his theory; but his polemical style is perhaps the main culprit. Mach seems to rely on the common sense of physicists, believing that their acquaintance with observation and experiment will teach them how to strike a balance between economy and comprehensiveness. He says little to them about how such balances are to be achieved; he prefers to assault them with slogans such as

> Physics is experience, arranged in economical order. (1943: 197)

Mach does, however, point to Copernicus, Galileo and Newton as having achieved great economies in their work; so perhaps he believes that only close study of historic masterpieces will reveal to present day scientists how they should proceed.

In the next section, we shall examine Mach's opposition to absolute space in the light of the preceding analysis of his phenomenalism and of his theory of economy, and we shall attempt to give a clear statement to 'Mach's principle'.

1.6 Who's afraid of absolute space?

We discovered in the preceding section that Mach singles out the concept of absolute space as being particularly heinous. But we also observed that Mach's phenomenalism is not a terribly secure basis for the repudiation of this concept. We said this for two reasons. First, his phenomenalism suffers from the grave difficulties mentioned which are peculiar to such sensationalistic philosophies of science. Secondly, it involves a commitment to the theory/observation dichtomy which few would now wish to maintain. Hence, I believe that Mach's phenomenalism *per se* is not sufficient to persuade us to abandon the concept of absolute space. But, more significantly, we are thereby given little reason even to *advocate* the repudiation of absolute space. Yet, following in Mach's footsteps, many physicists, for example Sciama and Dicke, have been advocates for Mach's cause.

My task in this section is to attempt to provide a philosophical justification for those physicists who have rallied to Mach's war cry against absolute space. As I have implied above, we still need to be armed with much more than Mach's phenomenalism; and we shall also have to take account of the very strong case against the theory/observation dichotomy.

Mach's antagonism towards absolute space is often supposed to be encapsulated in the celebrated 'Mach's principle' (MP). Einstein was the first to refer explicitly to MP in his paper 'Principle of general relativity' (1918). According to Einstein, MP states that

> (MP1) The inertial field is to be determined only by the distribution of mass-energy. (1918: 241)

And Einstein explains that

> I have chosen the name 'Mach's Principle' because the principle

> implies a generalisation of Mach's requirement according to which inertia should be reduced to the interaction of bodies. (1918: 241n)

Bradley, in Mach's *Philosophy of Science* (1971), tells us that 'there is no Mach's Principle in Mach's writings;' (1971: 145). Of course, Einstein's MP1, the first expression of Mach's principle as such, set as it is in a relativistic field context, is nowhere in Mach's books. But the general assertion that inertia has its source in material interactions is to be found in Mach's thought—as evidenced by the summary of §1.3 above. Indeed, we only need to remind ourselves of point (4) of that summary—namely, the inertial forces which we observe may be explained by the fact that they arise only in objects accelerating or rotating with respect to fixed celestial or material frames of reference—for us to see that this assertion is the essence of Einstein's MP1.

It is not difficult to remove MP from the relativistic context of MP1, and restate it as follows:

> (MP2) The inertial properties of material bodies are to be determined only by the distribution of material bodies.

In this form MP would be recognisable by, and acceptable to, Mach, given our discussion in §1.3. Rather than delve immediately into the intricacies of relativity, it may prove helpful to focus on MP2 as a statement of Mach's principle. Reinhardt, in his useful summary of the history of Mach's principle, maintains that we can state MP in the following general form:

> The inertial mass of a body is caused by its interactions with the other bodies in the universe. (1973: 529)

However, I prefer to use the neutral idea of 'determination' rather than that of 'causation' in such a general form of MP. Mach would certainly not have approved of the importation of causal language into a principle bearing his name. Mach makes his views quite clear in *The Science of Mechanics* when he says that

> In a lecture delivered in 1871, I outlined my epistemological point of view in natural science generally, and with special exactness for physics. The concept of cause is replaced there by the concept of function; the determining of the dependence of phenomena on one another, the economic exposition of actual facts, is proclaimed as the object (of science). (1960: 325)

Consequently, if we are to keep the Machian spirit alive in MP, I believe the notion of 'cause' should be avoided.

So, what is the status of MP2? Of course, just *how* the inertial properties are determined by the matter distribution is clearly an empirical matter as far as Mach is concerned: the size of particular inertial forces, the question of whether there is a reciprocal action of a body moving non-inertially upon the distant fixed stars, the problem of whether a 'mile-thick' bucket rotating around stationary water (with respect to the stars) produces inertial forces in the water, and so on—Mach indicates that these questions are empirical matters. But when we ask whether the inertial forces which we observe could be the result of anything other than material bodies, Mach's phenomenalism pushes us outside the empirical domain. Indeed, we would have to give MP2 an *a priori* status. For we would only be able to admit *material* bodies of the source of inertia: what cannot be perceived via sensory experience, for example absolute space, cannot be regarded as a source of any physical phenomenon. Such 'entities' as absolute space have no physical significance—this being conferred by the possibility of an entity being the subject of our sensory experience. Consequently, there would be no question of there being any empirical evidence for or against the essential thesis of MP that *material* bodies are the source of inertia. Again, we should note that there might be empirical evidence for or against some particular formulation of MP, be it MP1, MP2 or whatever. The action of bodies on bodies might vary inversely with distance, or with the inverse square of distance, or perhaps in a more complex way or, again, whether inertial forces are due to the *entire* matter distribution of just some *part* of the distribution—it is for empirical evidence to indicate the most appropriate characterisation of the relationship. What Mach's phenomenalism tries to force upon us is the conclusion that *only* material bodies are responsible for the inertial properties of matter. It is in this sense—as phenomenalism—that we would proclaim a principle like MP2 as *a priori*.

There are many versions of MP on offer similar, at least in spirit, to MP2. The following examples give us an indication of the variety of principles available, although each one reminds us of the essential idea of MP, namely that material bodies are the source of inertia.

(1) Sciama's version of MP:

> (MP3) Inertial motions are those that are unaccelerated relative to the 'fixed' stars—that is relative to some suitably defined average of all the matter in the universe. Moreover, matter has inertia only because there is other in the Universe. (1969: 19)

(2) Wheeler has a version which reads:

> (MP4) The geometry of spacetime and therefore the inertial properties

> of every infinitesimal test particle are determined by the distribution of energy and energy-flow throughout all space. (1964: 305)

(3) Raine tells us that the principle consists of the following ideas:

> (MP5) Only relative motion is observable, and hence there should be no dynamically privileged reference frames; . . . inertial forces should arise from a gravitational interaction between matter only; . . . spacetime is not an absolute element of physics, but its metric structure is totally dependent on the matter content of the universe. (1975: 507)

(4) Brans and Dicke offer us an equally lengthy version:

> (MP6) . . . the geometrical and inertial properties of space are meaningless for an empty space, . . . the physical properties of space have their origins in the matter contained therein, . . . the only meaningful motion of a particle is motion relative to other matter in the universe. (1961: 925)

We shall have to wait for Chapters 2 and 3 before many of the terms employed above can be made clearer. But the central theme of MP—namely, that *matter* (or mass–energy) is the source of inertia—is evident in each principle cited. But despite this consensus, there is some divergence of opinion, in particular concerning the following issues:

(i) which particular formulation should be given to MP;
(ii) whether MP should be used as a criterion of (*a*) theory choice or (*b*) selection of solutions of a given theory; and
(iii) whether or not the use of the central idea of MP to help us formulate either dynamical theories or solutions to given theories has *any* validity at all.

We shall consider questions (i) and (ii) in Chapter 3, when we have the tools of relativity theory at our disposal. Question (iii) is, of course, the focus of this section, concerned as it is with the *status* of MP. But we should note that as well as trying to provide some philosophical justification for MP, we must also show that the principle is neither theoretically nor empirically otiose. Once again, we shall have to wait until Chapter 3 before we are able to investigate the theoretical and empirical standing of MP. So here we shall concentrate on the philosophical status of MP. In looking at the status of MP, we will need to go over some of the ground covered in the last section. This is essential since the advocacy of MP seems more often than not to be based on some brand of phenomenalism.

A first response—and final for many—might be to say that MP has no convincing philosophical justification: such respondents might believe that MP derives any strength it possesses from Mach's phenomenalism, and they would therefore condemn MP on the grounds that phenomenalism does not provide us with a sound epistemology. But what would they say about physicists like Sciama, Dicke and Raine who seem to share Mach's antagonism to absolute space? If it is simply the case that these physicists oppose absolute space and its associated concepts on the grounds that space is 'unobservable', then they might be accused of philosophical naivety. For, unlike Mach, they have enjoyed the advantage of living through the demise of phenomenalism and positivism. They might also be accused of muddled thinking: why on earth should anyone wish to oppose a thoroughly respectable 'theoretical' term like 'space' when physics uses so many other, and perhaps less respectable, theoretical terms? Like Mach, physicists who advocate MP seem to desire the complete repudiation of absolute space. Yet, like Mach, they are tolerant of many other theoretical terms. So, without Mach's raw phenomenalism as an excuse, why have so many eminent physicists argued in favour of MP? Apart from complete naivety, I think that several explanations can be offered.

(1) Mathematicians might possess the intellectual drive to invent/discover new theories of new solutions—and they might see MP as a useful 'catalyst' for their mathematical ingenuities.

(2) Sophisticated empiricists might prefer to employ observational terms at the expense of theoretical terms whenever this is possible—and they might see in MP the chance to dispense with a high-level theoretical term without theoretical loss.

(3) Economisers might wish to avoid the use of unnecessary terms in physical theories—and they might see 'absolute space' as a paradigm case of a superfluous term.

Let us review the cases for each of these approaches.

(1) There are indeed some physicists who admit that they are inclined to view GTR and other theories as purely mathematical enterprises—for example Trautman who says that:

> One of the many unsolved problems connected with the general theory of relativity is whether the theory belongs to physics or rather to mathematics. One of my colleagues . . . said that those who work in the theory of relativity do so because of its mathematical beauty rather than because they want to make predictions which could be checked against experiment. I think there is some truth in this statement, and probably I am no exception to it. (1967: 7)

Whilst I have no doubt that some 'physicists' do regard GTR as almost a mathematical indulgence, I am inclined to think that those physicists who advocate MP believe that GTR is first and foremost a physical theory. Indeed, Sciama, Raine and others maintain that MP leads directly to empirical claims concerning the global structure of the universe (see Sciama (1959: 84) and Raine (1975: 507)). For them, mathematics is at the service of physics. But even if the mathematical approach provides a partial explanation for the advocacy of MP, it does not seem to give us sufficient reason to bind a physicist to MP rather than to some other, perhaps 'absolutist', principle. Hence, we must suspect those consistently in favour of MP of some other motivation, probably empirically based. This suspicion points us towards the second approach.

(2) Most present day physicists who advocate MP do not deny that their aversion to absolute space is based on the belief that space is not observable in any straightforward way; but they do *not* assert that observability alone should be the fundamental criterion according to which scientific terms may be accepted or rejected. Even Einstein, who once dismissed absolute space as

> a merely *factitious* cause, and not a thing that can be observed (1916: 113)

sometimes emphasises theory at the expense of observation and experiment. Referring to the equivalence of inertial and gravitational masses proposed by himself in 1907, Einstein says that he 'had no serious doubts about its strict validity, even without knowing the results of the admirable experiment of Eötvös'—theoretical considerations were the decisive factor; again, Einstein is reported to have said of Eddington's famous confirmation of GTR that *even without it* 'the theory is correct'.† More recent advocates of MP do not hesitate to employ high-level theoretical terms when the scientific context seems to demand it. Sciama, for example, mentions that singularities are a consequence of GTR (1971: 127). But it would take a bold heart to claim that singularities are clearly *observable* entities given our present knowledge. Like Mach, such physicists have to face a dilemma. Either:

(i) they must exclude all theoretical terms from physical discourse, however unpalatable this might seem to be, so that, even if they accept that all terms are theory-laden (in the sense that they depend

†These quotations are given in Holton (1973: 236–7); Holton quotes many such examples in trying to demonstrate Einstein's passage from phenomenalism to realism.

on the theory as a whole), no damage is done to their thesis that science deals only with observables; or

(ii) they must find a way of admitting the theoretical terms towards which they feel less antagonistic, yet still be able to exclude absolute space and other such disagreeable terms.

We saw in the preceding section that the first pathway out of the dilemma is unsatifactory—there is little cogency in the demand that science must deal only with observation. The second way is perhaps more promising for the advocates of MP. But exactly how they might escape the dilemma will depend on whether or not they decide to abandon the theory/observation distinction.

1.6(a) Away with the distinction

The physicists who advocate MP may well prefer to adopt a 'spectrum' view of scientific terms, such as that recommended by Maxwell (1962: 7) and others: a spectrum of terms and statements ranging from low-level 'observational' to high-level 'theoretical'. Hence, there would be no clear distinguishing mark of what is and what is not observational: a term would be either more or less 'observational'—but never purely observational, for all statements of the theory would depend upon the theory as a whole for their validity. The proponents of MP might then argue that the history of science has taught them to have far less confidence in the status of those terms nearer the high-level end of the spectrum: terms like 'quark', 'singularity' and so on. Their suspicious attitude towards the higher-level 'theoretical' terms enables them to retain Mach's positivistic bias in favour of observation and experiment, but they would also be able to drop the embarrassing commitment of positivists to a clearly definable set of observational terms and statements. In order to add weight to their argument, they might try to state their preference for 'observational' language in terms of 'personalist probabilities', perhaps following Hesse's Bayesian approach (Hesse 1974). But whatever framework they choose for their views, they would have no reason to be anything more than extremely suspicious about absolute space. There would be no *philosophical* grounds for them to *exclude* absolute space from scientific discourse. The lack of a firm observation/theory distinction prevents any *a priori* exclusion on the grounds that absolute space is at the 'theoretical' end of the observation/theory spectrum. For, if the validity of observation reports is guaranteed at least to some extent by the theoretical context as a whole—and if that context includes in some essential way absolute space—then they cannot dissociate the relevant observation reports from a theoretical commitment to absolute space. Hence, they cannot *exclude* absolute space from the

theoretical context *a priori*. Although some might hope that MP would involve the philosophical repudiation of absolute space, it is clear that this line of argument does not go quite so far. It expresses an empiricist prejudice against such high-level 'theoretical' notions and involves the hope that *a posteriori* evidence for their eventual exclusion will turn up someday, failing more meaty empirical evidence which might pull them further towards the observational end of the spectrum. But we should note that absolute space could only be excluded on this view when the evidence points towards a theoretical shift, with the resultant theoretical context having no essential commitment to absolute space.

1.6(b) Retain the distinction

Consequently, those who advocate MP might decide that a stronger empiricist line of argument is required: one which recognises that absolute space has a role in the theoretical context of dynamics, but which nevertheless excludes statements referring to absolute space from 'genuine' scientific discourse on the grounds that absolute space is not observable. However, so long as their arguments are based upon philosophical theories of perception, some data and some experience, they seem destined to run into difficulties similar to, even if not so acute as, those met by the earlier adherents to the belief that we can always distinguish the observational from the theoretical. I am certain that no clear distinction can be made between observational and theoretical language. I agree with van Fraassen that

> All our language is thoroughly theory-infected. If we could cleanse our language of theory-laden terms, beginning with the recently introduced ones like 'VHF receiver', continuing through 'mass' and 'impulse' to 'element' and so on into the prehistory of language formation, we would end up with nothing useful. The way we talk, and scientists talk, is guided by previously accepted theories. This is true also, as Duhem already emphasized, of experimental reports. Hygienic reconstructions of language such as the positivists envisaged are simply not on. (1980: 14).

However, van Fraassen also argues that despite the presence of theoretical terms in a given theory we can clearly distinguish those *things* and *events* which can be observed from those which cannot and we should *accept* scientific theories only insofar as what they say about the observable things and events are true. Hence, the commitment we have to theories is based upon the empirical adequacy of those theories:

> Science aims to give us theories which are empirically adequate and

> acceptance of a theory involves as belief only that it is empirically adequate; . . . a theory is empirically adequate exactly if what it says about the observable things and events in this world, is true—exactly if it 'saves the phenomena'. A little more precisely: such a theory has at least one model that all the actual phenomena fit inside. (1980: 12)

This anti-realist ploy is dubbed 'constructive empiricism' by van Fraassen. The assumption on which it is based is quite clear: that we can single out those things and events which are observable and distinguish them from those which are not. Van Fraassen believes that because this assumption is reasonable, we are able to accept scientific theories without committing ourselves to the existence of entities like micro-particles or absolute space.

The advocate of MP might be well pleased with constructive empiricism. Despite the fact that a theory might include a reference to absolute space, if space is not amongst the observables, then he has no reason to assert its existence, and his decision to accept the theory carries with it only the belief that the theory is empirically adequate. The part of the theory that matters is that concerned with observable things and events—the remainder is decidedly second-class. Certainly, the advocate of MP would be adopting a weaker position than Mach's, for absolute space is allowed a role in dynamics, but this role is merely instrumental according to van Fraassen. If it turns out that absolute space is not observable to the epistemic community, then van Fraassen permits us to exclude absolute space from the context which counts—that of empirical adequacy.

We must note, however, that van Fraassen seems to be employing a subterfuge here. He recognises that the theory/observation distinction is illusory so far as *language* is concerned. But he argues that a distinction can be made when we deal with *things* and *events*. He maintains that empirical investigation can reveal the extent of human capabilities as far as observation is concerned. We can then assert that a particular thing or event is observable:

> Science presents a picture of the world which is much richer in content that what the unaided eye discerns . . . science itself teaches us also that it is richer than the unaided eye *can* discern. For science itself delineates, as least to some extent, the observable parts of the world it describes. (1980: 58–9)

Van Fraassen also tells us that we can specify the limits of observation by characterising, for a given physical structure, exactly what of that structure is measurable/observable by the epistemic community.

In support of his argument van Fraassen presents a standard interpretation of the 'presence' of micro-particles in a cloud chamber:

> A look through a telescope at the moons of Jupiter seems to me a clear case of observation, since astronauts will no doubt be able to see them as well from close up. But the purported observation of micro-particles in a cloud chamber seems to me a clearly different case—if our theory about what happens there is right. The theory says that if a charged particle traverses a chamber filled with saturated vapour, some atoms in the neighbourhood of its path are ionized. If this vapour is decompressed, and hence becomes supersaturated, it condenses in droplets on the ions, thus marking the path of the particle. The resulting silver-grey line is similar (physically as well as in appearance) to the vapour trail left in the sky when a jet passes. Suppose I point to such a trail and say: "Look, there is a jet;" might you not say: "I see the vapour trail, but where is the jet?" Then I would answer: "Look just a bit ahead of the trail . . . there! Do you see it?" Now, in the case of the cloud chamber this response is not possible. So while the particle is detected by means of the cloud chamber, and the detection is based on observation, it is clearly not a case of the particle's being observed. (1980: 16–17)

Here van Fraassen is drawing a distinction between those things and events which can be observed under the appropriate conditions, and those putative entities and events which are 'observed' via their effects, i.e. between 'direct' and 'indirect' observation. We might be tempted to put the observation of absolute space in the latter category. The evidence for absolute space comes primarily from the existence of inertial forces which, if we accept Newton's argument, cannot be explained in material terms alone. Hence, because the 'observation' of absolute space can only be 'indirect', absolute space has no role in the *phenomenal* description of the world. That, van Fraassen contends, is the description which really matters.

In a discussion of realist and instrumentalist attitudes towards science, Fine (1986) points out that the instrumentalism evident in van Fraassen's work avoids the 'inflationary metaphysics' of realist approaches but suffers from its own kind of inflationism. An instrumentalist regards theoretical language as being merely instrumental, with no ontological significance. Hence the instrumentalist tries to 'deflate' the metaphysical commitments of scientific theories. But Fine argues that van Fraassen requires an epistemological distinction between belief and acceptance. This distinction inflates the *epistemology* of the scientist: although scientists may cut down on the number of entities in their theories, they are asked by van Fraassen to *believe* in some cases and merely to *accept* in others. The latter attitude is reserved for theoretical entities and statements. Is this a serious problem? Instrumentalists like van Fraassen may adopt the view that metaphysical inflation is somehow more heinous than epistemological inflation. This attitude might be acceptable if they

really do provide a clear-cut distinction between the observable and the theoretical. We would then have some powerful reasons to grant the distinction between belief and acceptance.

However, a question-mark hangs over van Fraassen's distinction between the observable and the unobservable—relying as it does on the doubtful assumption that we can *always* discover whether the objects and events referred to in a given theory are observable 'directly' or not. On van Fraassen's view the concept of observability is relative to the *beliefs* as well as the capabilities of the epistemic community. As the community changes, their belief about what is observable may also change.† Hence, a change in the theoretical context, and therefore in the beliefs of the community, might result in a previously 'unobservable' object becoming 'observable': for example, when a theoretical shift results in the identification of 'unobservable' genes with DNA molecules which are visible in micrographs.‡ Consequently, the assertion that an object or an event is observable seems to be grounded on shifting sand. Because we can and do change our minds, we can hardly claim that an object or event is observable once and for all, even given the stability of capabilities amongst the epistemic community. Even if we don't develop a new kind of vision, micro-particles could become visible to us, or, conversely, the moons of Jupiter might just disappear! All we need is an appropriate change in our theories. Whatever capabilities humans have with regard to observation, what we *say* about these observations will be constrained by our theories, both formal and informal, about the world. As Quine points out

> . . . it is misleading to speak of the empirical content of an individual statement; . . . Any statement can be held true come what may, if we make drastic enough adjustments elsewhere in the system. Even a statement very close to the periphery [of the theory's contact with experience] can be held true in the face of recalcitrant experience by pleading hallucination or by amending certain statements of the kind called logical laws. Conversely, by the same token, no statement is immune to revision. (1953: 43)

Herein lies a serious problem for van Fraassen. He maintains that scientists should accept theories when what they say about the observable world is true. But what they say about things and events—including the identification of things and events as observable—will be constrained by the theories which the scientists hold. The labelling of an entity as observable is theory-dependent. Yet

†See van Fraassen (1980: 18).

‡I am grateful to Hesse (1974: 29) for this example.

van Fraassen argues that scientists can isolate a model of a given theory which 'saves the phenomena'—despite the fact that the very isolation of the phenomena classified as observables must be theory-dependent. What is observable to the scientists can be discovered, according to van Fraassen, empirically. But he seems to forget that the methods and results of these empirical investigations will be constrained by the theories which the scientists hold. So, even if we allow the possibility of a 'pure observation', the characterisation of objects and events *as* observable must be theory-dependent.

I must conclude therefore that van Fraassen's attempt to revive the theory/observation distinction fails. Once he concedes—as I believe he must—that observational *language* is theory-dependent, he cannot proceed to draw a *clear* distinction between observable and unobservable *things* or *events*. For the very demarcation which he tries to make is itself theory-dependent. Consequently, we must discourage advocates of MP from following this path. There is no justification here for the claim that absolute space cannot be 'observed'. The inevitable upshot of the failure to proffer a convincing theory/observation distinction is that the advocate of MP who seeks to exclude absolute space from scientific discourse on phenomenal grounds alone is in a very weak position. As we found above, the best that he can do is to be deeply suspicious of high-level 'theoretical' language.

1.6(c) A third approach

The third approach open to those who advocate MP is to argue that the concept of absolute space is an unnecessary extravagance in dynamics. Like Mach, they might demand that our theories should be economical in content. They might then try to use Occam's razor to remove superfluous entities from the ontology of their theories. But, in order to rid themselves of absolute space on the grounds of economy, the physicisists who favour MP must show that no theoretical loss results from the repudiation of absolute space. However, if we accept Newton's two globes argument—and we have seen how powerful this seems to be—then there would indeed be a theoretical loss if we were to remove absolute space from Newtonian dynamics. For there would then be no immediate explanation for the presence of inertial forces in rotating bodies and systems. So physicists could hardly maintain that MP advocated on the grounds of economy gives us, by itself, sufficient reason to repudiate absolute space. Nevertheless, they might still maintain that MP should be advocated in the interests of economy. Whereas sophisticated empiricists might be suspicious of relatively high-level theoretical language; given that the history of science has provided good reason for this attitude, the economisers might follow Mach in maintaining that the economy of science is a fundamental scientific aim, for all the reasons seen in the preceding section. Hence

they might demand economy in dynamics if possible. Of course, they would need a little encouragement from the development of physics. There would have to be some promise of a theory which might make absolute space redundant. Otherwise their demand might seem to be merely wishful thinking. But it is clearly not irrational to commit oneself to the search for theories which are economical in content and form. And given the promise that GTR has provided for a theory without an essential commitment to an irreducible absolute space, the advocacy of MP on the grounds of economy also has much to commend it.

As we have seen, Fine (1986) accuses van Fraassen of multiplying epistemological attitudes. Does the demand for simplicity inevitably lead to epistemological extravagance? If we try to retain a clear-cut distinction between the observable and the unobservable, then two distinct attitudes are required: belief in what we observe, and mere acceptance of the theoretical. But the scientist who adopts the economising position above has no need to maintain a rigid distinction between observational and theoretical entities and statements. The 'spectrum' view of observation and theory may be adopted. Then those who search for simplicity in scientific theories may regard this spectrum with a corresponding spectrum of degrees of belief. The issue of mere acceptance is not raised. If an entity or statement *is* ruled out on grounds of economy, then zero degree of belief is accorded to that entity or statement.

What lessons can the advocate of MP learn from these remarks concerning the three approaches open to him?

(1) Treating dynamics as a mathematical game does not give him sufficient reason to repudiate absolute space.

(2) There are no *a priori* grounds available to justify the removal of absolute space from dynamics.

(3) His empiricist prejudices may only be voiced through distrust of higher-level 'theoretical' language.

(4) His prejudices may be fuelled by the belief that the history of science gives us every cause to distrust such language.

(5) He must seek out evidence, both empirical and theoretical, to support the repudiation of absolute space.

(6) Since the demand that scientific theories should be economical is not *prima facie* unreasonable, he may advocate the removal of absolute space from dynamics—if possible.

(7) Of course, he would be more likely to do this if he is a sophisticated empiricist who is suspicious of entities like absolute space.

(8) There must be some promise of a theory of dynamics which does not require absolute space in any essential and irreducible way.

Together these lessons bring us to the adoption of MP as a methodological injunction to rule out absolute space if possible. Hence physicists may advocate MP given their desire for economy in dynamics and their distrust of entities remote from the interface of theory with experience. But if they are to adopt MP in this spirit, they must recognise that the evidence may point overwhelmingly towards the retention of absolute space as an essential element of dynamics—despite their empiricist and relationist inclinations. Then they would be obliged to embrace the absolute. This is the unfortunate consequence of failing to discover convincing *a priori* grounds for the repudiation of absolute space.

Finally, we should note the kinship between the position reached in this section—a blend of empiricism and economy—and the views ascribed to Mach in the previous section. Although we have moved from a strong, *a priori* Machian relationism which incorporates the belief that economy is the fundamental aim of science, we have gained cogency by the formulation of a weaker, methodological relationism. Despite dropping Mach's embarrassing *a priori* prejudices, we have nevertheless retained a definite Machian flavour in a cogent justification for the advocacy of MP. On the one hand we have stressed the need for economy, and on the other hand we have kept a positivistic bias towards observation and experiment at the expense of the theoretical. The similarity between this more palatable relationism and Mach's gives support to my earlier claim that not all in Mach's philosophy of science is barren.

The search for simplicity leads us away from any *a priori* belief in science. No entity or statement is accorded any special status. The scientist is certainly driven by the conviction that observation and experiment are central features of science. Without these features, it would be hard to imagine what 'testing a prediction' might amount to. Science should not be a mathematical game devoid of testable content. Whatever else it involves, observation and experiment are *essential* features. But we do not need to believe any particular observation as an unquestionable fact. Nor should an experiment be seen as delivering any incontestable results. Hence, we leave behind not only the rigid positivism evident in some of Mach's thought but also the view that there might be *a priori* reasons for belief in any theory.

Albert Einstein (1922). Burndy Library, courtesy AIP Niels Bohr Library.

Chapter 2

The Foundations of the General Theory of Relativity

Introduction

We observed in Chapter 1 that the various formulations of MP produced by physicists, whilst containing references back to Mach's basic intuitions, go considerably beyond that idea that the interaction of mass upon mass is the source of inertial forces. This is inevitable given the fact that Mach's arguments are directed at a Newtonian context whereas MP is normally set firmly in a relativistic context. Einstein realises this, and says in his autobiographical notes

> Mach conjectures that, in a truly rational theory, inertia would have to depend upon the interaction of the masses, precisely as was true for Newton's other forces, a conception which for a long time I considered as in principle the correct one. It presupposes implicitly, however, that the basic theory should be of the general type of Newton's mechanics: masses and their interaction as the original concepts. The attempt at such a solution does not fit into a consistent field theory, as will be immediately recognised. (1969: 29)

When we move from classical to relativistic dynamics we meet the strange idea of a field—strange, that is, in the context of a theory of gravitation. As Einstein points out, Mach's 'Newtonian' viewpoint with matter interacting with other matter 'at a distance' is *prima facie* at variance with the field conception. Indeed, Einstein came to believe that Mach's prime virtue is not any positive programme for a relativistic dynamics but the negative dismissal of the Newtonian concept of absolute space. In a letter to his friend Besso in 1917, Einstein says of Mach's work

> It cannot give birth to anything living, it can only exterminate harmful vermin. (Holton 1973: 240)

Thus Einstein seems to disregard much that Mach says about the prospects for dynamics, but values Mach's dismissal of absolute space which is encapsulated in MP. However, we ought to point out in Mach's favour that he does consider the possibility of inertial forces arising *not* from the interaction of mass with mass, but from the interaction of mass with a '*medium*' in which the mass exists. Mach tells us that

> . . . in such a case we should have to substitute this medium for Newton's absolute space. (1960: 282)

Mach does not specify what the nature of such a medium might be; he could have in mind either some all-pervading ether or, indeed, some kind of gravitational field! Although the above remark is purely conjectural, we must grant that Mach's ideas on dynamics are much more extensive than some will acknowledge. Whether or not Mach in 1883 had the idea of such a field in mind, his remarks are certainly prophetic:

> . . . although, practically, and at present, nothing is to be accomplished with this conception, we might still hope to learn more in future concerning this hypothetical medium; and from the point of view of science it would be in every respect a more valuable acquisition than the forlorn idea of absolute space. (1960: 283)

Nevertheless, Mach's stubborn reluctance to move even a short way from what he conceives of as the realm of observation and experiment is at least a partial justification of Einstein's verdict. For Mach reminds us almost immediately that the science of his day was only competent to deal with material bodies and, therefore,

> It will be found expedient provisionally to regard all motions as determined by those bodies. (1960: 283)

Einstein's inclination to discount Mach's teaching comes not just from the acknowledgment that Mach's ideas on dynamics are set in a classical context, but perhaps also from Einstein's movement away from phenomenalism. Indeed, this movement is underlined by Einstein's occasional emphasis of theoretical considerations at the expense of observation. Holton has dubbed Einstein's 'progress' as a 'pilgrimage' towards a 'rationalistic realism' (1973: 240). Certainly, Einstein's views about developments during the early years of

relativity theory might well have shocked Mach, the mentor of his youth.

I agree with Holton that Einstein was hardly behaving like a phenomenalist during his years of 'pilgrimage'. Indeed, Einstein does say that his views on science owe much more to Hume than to Mach (see the autobiographical notes (1969: 53) and his correspondence with Besso (Holton 1973: 232)). But I cannot accept Holton's rather stark contrast between Mach and Einstein. Holton claims that

> To [Mach] the fundamental task of science was economic and descriptive; to [Einstein] it was speculative, constructive and intuitive. (1973: 239)

I have no reason to quarrel with Holton's implication that Mach fails to give any credit to the speculative, theoretical approach to science. For, as we have seen, this is quite clearly the case. Indeed, Einstein makes this point in his autobiographical notes:

> I see Mach's greatness in his incorruptible scepticism and independence; in my younger years, however, Mach's epistemological position also influenced me very greatly, a position which today appears to me to be essentially untenable. For he did not place in the correct light the essentially constructive and speculative nature of thought and more especially of scientific thought; in consequence of which he condemned theory on precisely those points where its constructive–speculative character unconcealably comes to light, as for example in the kinetic atomic theory. (1969: 21).

But Holton also suggests that economy is at most of secondary importance in Einstein's work. Yet both scientists, albeit in their different ways, show the goal of economy in science to be a principal concern. We have examined Mach's opinions on this subject. In the rest of this chapter, we shall discover how considerations of simplicity and economy are involved in Einstein's progress towards the field equations of GTR.

2.1 The principles of equivalence and covariance

Perhaps the first important landmark in setting up GTR came with Einstein's realisation that inertial and uniform gravitational forces seem to be observationally indistinguishable. Einstein bases his ideas on Galileo's discovery that all bodies undergo the same acceleration in a given gravitational field, thus demonstrating a proportionality between gravitational mass and inertial mass. Einstein tries to show

his point, namely that inertial and gravitational forces are equivalent, with his 'man-in-a-chest' thought experiment which I shall briefly describe (see Einstein (1954: 66–70)).

A man in a chest believes himself to be suspended in a gravitational field; he comes to this conclusion when he realises, as did Galileo, that, whatever object he releases, the acceleration of the object towards the floor of the chest is always of the same magnitude. However, the man's belief is mistaken. In fact, the chest is being uniformly accelerated in a direction opposite to that of the supposed gravitational field: this acceleration taking place in 'empty space', far from other material bodies. Now, Einstein asks, ought we to smile at this man's belief that he is at rest in a gravitational field? Not at all, for even though the chest is being accelerated relative to the 'empty space' frame of reference, the behaviour of his environment fulfills all the criteria of being at rest in a gravitational field. That is, there is no difference from this man's point of view between a gravitational force and an inertial force acting upon his environment: the two forces seem to be physically equivalent.

In this account we can observe that Einstein makes a jump from the proportionality of gravitational and inertial masses to the indistinguishability of gravitational and inertial forces. The man *believes* that any object released by him falls according to Newton's second law of motion:

$$F = ma \tag{2.1}$$

where the mass is *gravitational*. But Einstein tells us that the mass here is really *inertial*. If the man has no way of telling whether the chest is accelerating in a gravitational field or relative to an inertial frame, and for this to be so we must suppose that gravitational mass equals inertial mass (or else the objects would behave differently in the two situations), then Einstein concludes that, from the man's point of view, the force due to the gravitational acceleration is equivalent to the force experienced under 'ordinary' acceleration.†

Einstein is advocating *two* equivalences in his short story:

(i) the equivalence of inertial and gravitational masses; and
(ii) the equivalence of inertial and gravitational forces.

Equivalence (i) is concerned with material bodies, i.e. with particles etc that possess some rest mass. This equivalence does not seem to be problematic. Its Galilean pedigree and its high empirical standing

†The supposition that inertial and gravitational masses are proportional, and can be made equal by a suitable choice of units, has received a great deal of experimental confirmation, most notably by Eötvös in 1889 and 1922.

seem to guarantee its validity. Consequently, Einstein seems justified in arguing that the essentially Newtonian distinction between gravitational and inertial masses should be dropped. Einstein and his collaborator Grossmann tell us that equivalence (i) should be regarded as

> . . . an exact law of nature that must be expressed as a foundational principle of theoretical physics. (Einstein 1913: 225)

Sachs, however, maintains that we cannot assert this equivalence as such a foundational principle of GTR. He says this because gravitational mass remains undefined in GTR. In Newtonian theory, this mass was given by the law of gravitation

$$F = \frac{Gmm'}{r^2} \times a. \tag{2.2†}$$

However, this action-at-a-distance law of gravitation is not valid in GTR, except as a 'limit' in certain conditions (low gravity, low speeds and small spatial separations). For this reason Sachs tells us that we could not incorporate any law expressing the *general* equivalence of inertial and gravitational masses in GTR. The best we can do is to see in equivalence (i) a 'principle of correspondence' which shows the meeting points of GTR and Newtonian theory. On this basis, Sachs contends that

> . . . it is logically false to assert the (weak form of the) equivalence principle as a bona fide underlying axiom of the theory of general relativity. (1976: 226)

If this 'weak principle of equivalence' is exactly what Sachs says it is, namely the assertion of equivalence (i), then I do not hesitate to agree with him. However, in formulating GTR Einstein uses not so much equivalence (i), but the more general consequence of accepting this equivalence as a physical fact. If all material particles behave in the same way in a given gravitational field, then we can safely infer that physics has only one kind of mass to deal with. It is this inference that is the basis of the 'weak principle of equivalence'; and this principle is, as Anderson (1967: p.337) says, 'crucial to general relativity'.

Whatever particle we release in a given gravitational field will obey the relativistic equation of motion:

$$\frac{\mathrm{d}^2x^\mu}{\mathrm{d}\lambda^2} + \Gamma^\mu_{\rho\sigma}\frac{\mathrm{d}x^\rho}{\mathrm{d}\lambda}\frac{\mathrm{d}x^\sigma}{\mathrm{d}\lambda} = 0 \tag{2.3}$$

†Note that the force in equation (2.1) is more general than that in (2.2) which is gravitational only.

for a suitable choice of the affine 'path' parameter.† A free-falling particle will follow a geodesic, i.e. the straightest of all possible curves in space–time. Our observation of any such particle will always reveal the same affine connection Γ in any given region of space–time.‡ This connection depends upon the curvature of space–time; hence the paths of our test particles will be curved rather than the straight Euclidean lines found in Newtonian theory. Consequently, we may regard the affine connection as a field which determines the paths of test particles at all points of space–time.

This geometrical preamble leads us to what is *usually* regarded as the 'weak' principle of equivalence. Misner, Thorne and Wheeler state it as follows:

> The path through spacetime of a freely falling, neutral test body is independent of its structure and composition. (1973: 244)

If all we are concerned with are test particles that have some positive rest mass, then we can say that each will behave as any other particle in any given gravitational field. The mass, of course, will not be constant; in the Special Theory of Relativity (STR) Einstein demonstrates that mass varies with velocity. But in this relativistic statement of the equivalence principle we have no need for any distinction between gravitational and inertial masses. The *fact* of equivalence of all masses allowed Einstein to make the first step in the construction of GTR: by treating all particles alike we can represent the motion of particles simply, using the 'geometrical' concept of the affine connection.

We should note that Misner *et al*'s statement of the weak principle of equivalence is wider in its scope than equivalence (i), by itself, can allow. For it applies to all particles including those with zero rest mass. Why is this? If we remember the man in Einstein's chest, then we might well offer him light sources etc to help him in his quest to distinguish the two kinds of motion: inertial and gravitational. If light followed one path under the influence of an inertial force and a different path under a gravitational force, we would then have a certain experimental method for finding what sort of force is acting on released particles etc. To avoid this difficulty, and to widen the scope

†We should note that this equation makes no reference to mass: this indicates that GTR simply assumes that only one kind of mass is to be dealt with in its equations. The affine 'path' parameter can exist even in the absence of a metric despite the fact that the Γ function is a derivative of the metric tensor (see Sklar (1974: 49–50) and Weinberg (1972: 73–7)).

‡Provided that we neglect any influence the particle may have on the sources of the field (see Anderson (1967: 335)).

of the idea of equivalence, Einstein postulates that *all* particles do behave alike regardless of the nature of the motion. Hence, light particles too must follow geodesics, *null* geodesics, which, under the influence of the gravitational field, would 'bend' with the field. Einstein makes this prediction in his 'Influence of gravitation on the propagation of light' (1911).

The second equivalence (ii) has given rise to a stronger principle of equivalence. This weak/strong distinction although implicit in Einstein's work is made clear by Dicke who says

> By the weak principle (of equivalence), I mean that, up to the great accuracy of the Eötvös experiment, all bodies move along the same kind of geodesic paths . . . The strong principle says something more than this. It says that, in a free-falling laboratory, if one does experiments locally, one observes the same laws of physics—including all the numerical content that one observes anyplace else, including any gravity-free place. *This is the principle that is the basis of general relativity*. (1964: 13, my emphasis)

Now, Dicke's strong principle of equivalence which refers to 'local' experiments is the result of considering certain problems concerning Einstein's equivalence (ii). The 'man-in-a-chest' example makes a number of simplifying assumptions. When we drop these we move towards Dicke's principle. Einstein wants us to accept that all objects do behave alike regardless of whether the force is inertial or gravitational. If the two kinds of force are observationally indistinguishable, then we must be able to carry out all kinds of physical experiments in the chest and still be unable to find any distinguishing feature of either force. The consequence of such an equivalence is that just as 'inertial' forces can be 'transformed away' by changing the frame of reference from which we measure them, the same must be true for 'gravitational' forces; there is just *one* kind of force which we can treat in exactly the same manner whether we *call* it inertial or gravitational. But we know that gravitational forces are, in general, quite different from the inertial force in the 'man-in-a-chest' experiment. Consider the gravitational field about the Earth: it is directed more or less radially inwards. So, two objects falling together in this field would approach each other. If a man in the chest dropped two objects distant x apart from a height h, he could measure the distance between the objects at height h' at a later time. If the distance here were to be less than x, then the man could rightly say that the chest was experiencing the effects of a gravitational field. If the distance remained constant, then he could tell us that the chest was accelerating in distant space. Moreover, we would also have to take into account the fact that, in the gravitational case, the various

particles of each object released would approach each other. Hence all he would need would be two test particles to enable him to distinguish between inertial and gravitational forces.

We can now see why Dicke insists that experiments should be done *locally*. We can only treat inertial and gravitational forces alike if we take a sufficiently small portion of space–time around the particle in which we are interested. Hence, we must move from global to local reference frames. As we saw in Chapter 1, Newton and Mach both stated their allegiance to global inertial reference frames: for Newton the frame was absolute space, and for Mach it was the fixed stars (except perhaps when we talked of the possibility of a pervading medium). For Einstein, however, local inertial frames are the only ones which need to be employed. By working with these reference frames we can always transform away forces. Misner, Thorne and Wheeler state their advantages succinctly:

> What is direct and simple and meaningful, according to Einstein, is the geometry in every local inertial reference frame. There every particle moves in a straight line with uniform velocity . . . Collision and disintegration processes follow the laws of conservation of momentum and energy of special relativity. (1973: 19)

So, when we consider space–time locally, the laws of special relativity seem to hold good, and departures from inertial motion can now be determined by reference to these local inertial frames, or 'Galilean regions' as Einstein sometimes calls them (1956: 57). These regions correspond to the free-falling laboratories of Dicke's principle: in both cases inertial and gravitational forces have been transformed away.

The geometry of these regions is given by the expression $ds^2 = g_{ik}dx^i dx^k$ where the invariant ds^2 gives the separation ds between two neighbouring points in space–time. The metric tensor g_{ik} is a function of the point at which it is evaluated, and in local inertial frames g takes on the constant values

$$\begin{matrix} 1 & 0 & 0 & 0 \\ 0 & -1 & 0 & 0 \\ 0 & 0 & -1 & 0 \\ 0 & 0 & 0 & -1 \end{matrix}$$

This simple form of the metric is the Minkowskian metric η_{ik} of special relativity.

The conception of local inertial frames takes us into the centre of GTR. Einstein uses these frames to take apart the complex overall view of space–time. But they only suffice for our non-gravitational calculations—they tell us nothing, at least directly, about gravity. Einstein realised that gravitation could be manifested in the way that

our local 'pictures' of space–time are fitted back together to restore our global view. This reconstruction is of the gravitational field, which is manifested in the curvature of space–time. Our free test particles fall in straight lines *locally*; with the global view of the gravitational field these lines are curved, as the particles follow the geodesics which the geometry of space–time forces upon them. We shall see in §3.1 that the curvature of space–time is related to the distribution of masses in the space–time via the field equations.

Will (1979, 1981) presents a helpful analysis of equivalence which enables us to clarify its role. He distinguishes three principles.

(1) The weak equivalence principle (WEP):

> if an uncharged test body is placed at an initial event in spacetime and given an initial velocity there, then its subsequent trajectory will be independent of its internal structure and composition. (1981: 22)

(2) The Einstein equivalence principle (EEP):

> (i) WEP is valid, (ii) the outcome of any local nongravitational test experiment is independent of the velocity of the (freely falling) apparatus, and (iii) the outcome of any local nongravitational test experiment is independent of where and when in the universe it is performed. (1981: 22)

(3) The strong equivalence principle (SEP):

> (i) WEP is valid for self-gravitating bodies as well as for test bodies, (ii) the outcome of any local test experiment is independent of the velocity of the (freely falling) apparatus, and (iii) the outcome of any local test experiment is independent of where and when in the universe it is performed. (1981: 82)

If a theory of gravitation accords with EEP, then it must be a *metric* theory. This is clear from the discussion above and from an inspection of Will's EEP, for EEP will lead us directly to the idea of a local inertial frame. Hence any metric theory of gravitation will obey EEP (and therefore WEP).† But there are many metric theories other than GTR, as we shall discover. Accordingly, Will points to SEP as a prerequisite for GTR. Clearly, if SEP is valid, so too are WEP and EEP, since these are special cases of SEP. The strong equivalence principle extends the range of equivalence to self-gravitating bodies and gravitational forces. Will subsequently points out that GTR satisfies SEP: the metric g seems to be isolated by this principle as the only possible

†We should note also Schiff's conjecture: that any complete, self-consistent theory of gravity which embodies WEP necessarily embodies EEP (see Will (1981: 38–45) who gives compelling evidence for the truth of this conjecture).

gravitational field.† Hence theories with additional gravitational fields, whether or not these depend on the metric field g, are excluded. Several theories, e.g. the Brans–Dicke 'scalar field' theory and Rosen's bimetric theory, do not satisfy SEP because they postulate additional fields.‡ But Will leaves open the question of whether or not a theory other than GTR might satisfy SEP.

What is clear from the discussion above is that strong equivalence demands a theory which: is a metric theory incorporating the idea of a local inertial frame; in the 'local' limit, the non-gravitational laws of physics must be those of STR; and the metric should be given by the tensor g_{ik}.

Although the strong principle of equivalence *appears* to have taken us to the threshold of the field equations, a number of writers have questioned the validity and importance of the principle for GTR (see, for example, Synge (1966), Anderson (1967) and Sachs (1976)). Synge maintains that the 'principle' was only a guide for Einstein in his development of GTR. He says

> The Principle of Equivalence performed the essential office at the birth of general relativity but . . . I suggest that the midwife be now buried with appropriate honours. (1966: ix–x)

The 'office' which equivalence did perform is not in dispute: it formed a link between STR and GTR helping to *guide* Einstein towards the field equations. Dicke has told us that the strong principle is the basis of GTR. No one denies its *historical* role; but is equivalence anything more than a guiding principle which Einstein once used and can now be ditched? Zahar (1977) speaks for Synge and many others when he says that the strong principle of equivalence has only heuristic value for GTR. But do we have to concede that the strong principle has no place *within* GTR? That is, does the principle tell us something about gravitation and the laws of GTR rather than something about how Einstein happened to arrive at a theory of gravitation in 1915/16? Zahar and others who agree with his position maintain that the strong principle is not *strictly* valid in GTR, and therefore must be denied any full, logical status in the theory. Before assessing their claims in §2.2, I propose to examine Reichenbach's view of the principle of equivalence. Reichenbach's account is important for two reasons. First, it gives us some evidence for the assertion that this principle is the 'basis' of GTR. Secondly, it introduces us to the claim that equivalence has an essentially Machian character. Reichenbach's account of equivalence, in *The Philosophy of Space and Time* (1958) was first

†Will admits that no rigorous proof of this is available (1981: 83).

‡These theories will be discussed in Chapter 5.

published in 1927, just a few years after the completion of GTR.

Reichenbach maintains a view of physics and its role which has a distinct Machian character. Since much that Reichenbach has to say about equivalence is affected by his general views, a brief account of these will enable us to clarify his more particular claims. His general ideas are embodied in his 'theory of equivalent descriptions'; and he maintains that the concepts of dynamic relativity incorporate the distinction which this theory brings to the forefront. Reichenbach tells us that we must distinguish the metrical field from the gravitational field. The former is represented by the tensor function g as a whole (by which I believe Reichenbach means g abstracted from any components). The latter is represented by $g_{\mu\nu}$—the particular set of components involved in a given quantitative representation of the field. When we change our choice of coordinate systems, we do not change the character of the metrical field tensor g; it is invariant. What we do alter is the set of components of the tensor, e.g. we move from $g_{\mu\nu}$ to $g_{\mu'\nu'}$. Thus,

> The system of the tensor components is covariant, i.e. it has a different numerical composition for each co-ordinate system. (1958: 236)

The upshot of this covariant character of the gravitational field is that, by a suitable choice of coordinate systems, we can transform away the field (locally, of course). Reichenbach maintains that each tensor function $g_{\mu\nu}$ together with its coordinate system provides an equivalent description of the invariant physical world. And, this, Reichenbach claims, 'expresses a basic idea of modern science' (1958: 236). What Reichenbach is getting at here is made clear in a statement which he makes at the end of *The Philosophy of Space and Time*:

> Mathematical space is a *conceptual structure*, and as such *ideal*. Physics has the task of co-ordinating one of these mathematical structures to *reality*. In fulfilling this task, physics makes statements about reality, and it has been our aim to free the objective core of these assertions from the subjective additions introduced through the arbitrariness in the choice of descriptions. (1958: 287)

So, we might have competing theories making equivalent descriptions of the same phenomenon, or, as in the case of GTR, competing 'views' of the field being (generally covariant) descriptions of an (invariant) reality. Thus, we should see the metrical field as an instance of the objective reality to which physics addresses itself via the 'subjective' descriptions of particular gravitational fields.

Why do I think that all this has a Machian flavour? Reichenbach certainly does not make any such claim. Mach, however, makes two

assertions which seem to emerge in Reichenbach's account. First, Mach says that

> The universe is not *twice* given, with an earth at rest and an earth in motion; but only *once* with its *relative* motions alone determinable. (1960: 284)

and, secondly, that

> We may interpret the one case that is given us in different ways. If, however, we so interpret it that we come into conflict with experience, our interpretation is simply wrong. (1960: 284)

Mach's world view here is of a factual universe, a physical reality, which physics may interpret in various ways, successfully or otherwise. For example, Mach sees the Ptolemaic and the Copernican views as being 'equally correct' descriptions of the physical world. All this points to the fact that Reichenbach and Mach share the thesis that physics can provide us with 'equivalent' or 'equally correct' descriptions or intepretations of a single reality.

Reichenbach believes that Mach's principle may be given two senses: one philosophical, the other physical. The first epistemological sense commits us to the relativistic view that

> Every phenomenon is to receive the same interpretation from any moving co-ordinate system

and the second gives expression to the empirical content of the principle:

> All physical phenomena depend only on the relative position of bodies and not on the positions of these bodies in space. Two similar systems differently oriented in space must therefore show the same physical phenomena. (1958: 217)

We can trace these dicta back to Mach's remarks on dynamics, and particularly to those concerning Newton's rotating bucket thought experiment. But Reichenbach places stress on Mach's conjectures, rather than upon what Mach accepted as factual. That is, he emphasises Mach's contention that we might conceive the principles of mechanics so that 'even for relative rotations centrifugal forces arise' (1960: 284). However, Reichenbach fails to mention that Mach's ideas were conjectural in this context. Reichenbach then goes on to say that 'Mach's claim' for a *general* dynamic relativity is 'retained in the modern theory of relativity' (1958: 215). It is the

substance of this claim which appears as the second sense of Reichenbach's Machian principle; it is an empirical claim with observational consequences: we ought to be able to measure the effects of a thick-sided bucket rotating relative to its water content and to the 'fixed' stars. Reichenbach adds that Mach's claim is incorporated in Einstein's principle of equivalence. Reichenbach says that

> The equivalence of inertia and gravity is the strict formulation of Mach's principle in the (physical) sense. It implies that every phenomenon of inertia observable in an accelerated system can also be explained as a gravitational phenomenon; therefore it cannot be interpreted to indicate uniquely a state of motion. (1958: 225)

It is, of course, this equivalence which is the basis of the strong principle of equivalence. How does Reichenbach justify his assertion that Mach's views of dynamics are embodied within the principle of equivalence? The consequence of the general dynamic relativity outlined in the thick-sided bucket experiment is this: we can explain the bucket experiment in *two* ways—either the bucket is rotating relative to the fixed stars with inertial forces in the bucket or the bucket is fixed and the stars are rotating, thus producing (by some unspecified mechanism) gravitational forces in the bucket. The physical effects, according to Reichenbach, are the same in both cases. Hence we can maintain, with Mach, that they are dynamically equivalent. Reichenbach then compares this with one of the standard examples used to demonstrate the principle of equivalence; he cites the example of a 'freely falling elevator . . . in which the gravitation of the earth is transformed away' (1958: 225). As Einstein shows in his man-in-a-chest experiment, if we were to halt the elevator, an observer within would be unable to say whether or not the subsequent behaviour of a test system would be caused by inertial or gravitational effects. So, Reichenbach tells us that Mach's 'claims' for (or, we should say, conjectures about) the bucket experiment and Einstein's claims for freely falling boxes etc both establish the equivalence of inertial and gravitational forces.

Reichenbach then goes on to formulate a principle of local inertial systems on the basis of the idea of equivalence. This principle is essentially the same as Dicke's strong principle, and like Dicke he maintains that it is at the core of the general theory. Reichenbach maintains that equivalence's prediction, that light 'bends' under the influence of gravitation, compels us to move away from STR:

> It is not the astronomical inertial systems, but the local inertial systems, for which the special theory of relativity holds . . . We may

> thus speak of the principle of local inertial systems, which states that the local inertial systems are those systems in which the light and matter axioms [i.e. laws of STR] are satisfied. (1958: 230)†

Throughout his account of equivalence Reichenbach relies on his views of a single physical reality which may be given different but still equivalent interpretations. We have seen that he believes this view to be true to Mach's thought and contained within GTR. Whether the principle of equivalence is Machian or not, is a question we shall examine in §2.3. The rest of this section will be concerned, first, with the claim that equivalence is at the heart of GTR and, secondly, with the view that the principle of general covariance is a foundational principle of GTR. We shall see that the strong principle of equivalence leads us fairly naturally to the principle of general covariance. One of our problems will be to see exactly how these two principles are related.

The strong principle requires that the non-gravitational laws of STR hold good in the 'flat' space–time of local inertial frames, which corresponds to the space–time of special relativity. How can we reformulate the laws of STR so that they are valid in curved space–time? As Misner *et al* ask, how can we 'mesh together' the laws of STR with a field theory of gravitation? This trio of physicists gives an example which demonstrates the idea that Einstein employs.‡ The example exploits, as Einstein does, the transformational properties of tensors.

Consider the law of local energy-momentum conservation; in flat space–time the law states:

$$\nabla \cdot \boldsymbol{T} = 0. \tag{2.4}$$

We can rewrite the abstract geometric form of the law§ in component form for a global Lorentz frame of flat space–time:

$$T^{\mu\nu}{}_{,\nu} = 0 \tag{2.5}$$

where the comma signifies ordinary partial differentiation. According to the strong principle of equivalence this law is applicable at the origin of a local inertial frame of curved space–time:

$$T^{\hat{\mu}\hat{\nu}}{}_{,\hat{\nu}} = 0. \tag{2.6}$$

†Reichenbach adds a footnote to this passage, which, as we shall see later, turns out to be important. He tells us that 'strictly speaking' we should say '. . . are satisfied to a higher degree of approximation' (1958: 230). How the idea of local inertial frames is involved in the curved space–times of GTR is exactly the issue which Zahar, Sachs and others raise: *they* also say 'strictly speaking' the idea is only an approximation.

‡See Misner *et al* (1973: 385–7).

§See Misner *et al* (1973: 384–7).

The 'hats' here signify local components. Now, since the affine connection coefficients vanish at the origin of this frame, the covariant derivative reduces to the ordinary derivative.† Taking advantage of this we can rewrite (2.6) in covariant form:

$$T^{\hat{\mu}\hat{\nu}}{}_{;\hat{\nu}} = 0 \tag{2.7}$$

where the semicolon signifies covariant differentiation. These are the local components of the abstract geometric law in curved space–time:

$$\boldsymbol{\nabla}\cdot\boldsymbol{T} = 0 \tag{2.8}$$

whose component formulation in any reference frame is:

$$T^{\mu\nu}{}_{;\nu} = 0. \tag{2.9}$$

It will be immediately clear that equations (2.4) and (2.8) are identical; and that (2.5) and (2.9), their respective component formulations, differ only in the method of differentiation. Misner, Thorne and Wheeler conclude this example by saying that the reformulation of the laws of STR turns out to be a 'simple' matter:

> The laws of physics written in component (tensor) form, change on passage from flat spacetime to curved spacetime by a mere replacement of commas by semicolons. (1973: 387)

We should note that this example stipulates that a law of STR should hold *at the origin* of a local inertial frame. Up until now we have talked rather loosely, of the laws holding 'locally'; but, strictly speaking, in the arbitrarily curved space–time of GTR we cannot be so vague. That a law of STR holds in some *finite* region of curved space–time can only be, at best, a good approximation. If it holds at all in curved space–time, a law must do so at the origin of a local inertial frame, i.e. at a single point-event. The consequence of this is that the strong principle must concern itself only with the behaviour of test particles, i.e. point masses, if it is to be considered exact.‡

Misner *et al*'s example also introduces us to the idea of an equation being generally covariant. Just why Einstein employs this idea in GTR will require some background information. Einstein was convinced

†For some details of this see Misner *et al* or Adler *et al* (1975: 69).

‡Two difficulties arise for this. First, Anderson has pointed out that even for particles that can be considered as point masses, the strong principle will not apply to those which possess non-inertial properties such as spin, for this would be present, contrary to the principle, at the origin of the frame (1967: 337). Secondly, Sachs has argued that point masses cannot be considered separately from the field in GTR: but the strong principle does treat particles in this way and therefore makes an approximation which, although it may be a good one, betrays the inexact nature of the strong principle (1976: 228). These problems will turn out to be irrelevant when we resolve the question of the status of the principle.

that the special postulate of relativity (or, as it is sometimes called, the restricted principle of relativity) was inadequate. According to Einstein, this postulate demands that

> If, relative to (an inertially moving system) K, K′ is a uniformly moving co-ordinate system devoid of rotation, then natural phenomena run their course with respect to K′ according to exactly the same general laws as with respect to K. (1960: 13)

Of course, STR is consistent with this demand: its laws hold good in all such systems. However, the postulate gives a preferred status to inertial frames of reference, and this was part of the cause of Einstein's dissatisfaction with STR. Einstein says

> . . . to Mach's question: "how does it come about that inertial systems are physically distinguished above all other co-ordinate systems?" this theory [i.e. STR] offers no answer. (1969: 63)

That is, STR does not offer an epistemologically satisfactory answer to Mach's question. For STR, just as Newtonian mechanics, singles out systems in uniform motion for special treatment: for these systems do not accelerate or rotate with respect to space. Consequently STR endows space–time with a structure that cannot be reduced to the objects within it. Inertial forces will arise in those bodies which depart from inertial motion with respect to that structure, and the natural paths of bodies which are free of forces will be determined by that structure. This Einstein found distasteful; the 'privileged' space of STR and of Newtonian mechanics

> . . . is a merely *factitious* cause, and not a thing that can be observed. (Einstein *et al* 1923: 113)

Here Mach's influence is very clear indeed.

These problems with STR prompted Einstein to put his faith in a generalisation of the special postulate. His *general* postulate of relativity (or general principle of relativity) requires that

> The laws of physics must be of such a nature that they apply to systems of reference in any kind of motion. (Einstein *et al* 1923: 113)

But this postulate must be seen as a statement of conviction, as the preface to a manifesto rather than as the substance of a manifesto for a general theory of dynamics. In Einstein's paper of 1916 much of this substance is provided; Einstein maintains that a physical theory will satisfy the general postulate if it satisfies the rather more specific general principle of covariance. This principle requires that

> The general laws of nature are to be expressed by equations which hold good for all systems of co-ordinates, that is, are co-variant with respect to any substitutions whatever (generally covariant). (Einstein *et al* 1923: 117)

We have seen in Misner *et al*'s example an equation which satisfies this principle. An equation is generally covariant if it preserves its form under general coordinate transformations. The example shows that by using tensors the form of the equation $\boldsymbol{\nabla} \cdot \boldsymbol{T} = 0$ is preserved. The principle of general covariance therefore implies that if a law holds good in one coordinate system, it will hold good in all such systems.

Before I try to establish what the status of the principles of strong equivalence and general covariance is in GTR, I shall examine the relationship between the two principles. I believe that by considering what seems to be an erroneous restatement of the principle of general covariance this relationship will be made clear. In *Gravitation and Cosmology* Weinberg says that the principle of general covariance states that

> . . . a physical equation holds in a general gravitational field if two conditions are met:
>
> 1. The equation holds in the absence of gravitation; that is, it agrees with the laws of special relativity when the metric tensor $g_{\alpha\beta}$ equals the Minkowski tensor $\eta_{\alpha\beta}$ and when the affine connection $\Gamma^{\alpha}_{\beta\gamma}$ vanishes.
> 2. The equation is generally covariant; that is, it preserves its form under a general co-ordinate transformation $x \rightarrow x'$.
>
> (1972: 92-2)

According to Weinberg this principle *follows from* the principle of equivalence. He has previously stated the equivalence principle in essentially the same way as Dicke's strong principle, i.e. both say that we can always construct local inertial frames in which the laws of STR hold (1972: 68). It is immediately evident that Weinberg's condition (2) above is also essentially the same as Einstein's statement of the entire principle of general covariance. The introductory words before Weinberg's two conditions are understood in Einstein's principle, and emerge from the context of that principle. The only significant difference is therefore Weinberg's condition (1). Now, condition (1) tells us that the laws of physics should hold at the origin of a local inertial frame, for it is at this point that $g_{\alpha\beta} = \eta_{\alpha\beta}$ and that $\Gamma^{\alpha}_{\beta\gamma} = 0$. But this is surely part of the strong principle. The strong principle can be split into two parts:

(i) we can always construct local inertial frames at any point in

space–time; and

(ii) at the origin of these systems the laws of STR hold.

It is clear that (ii) depends upon the possibility of realising (i). It is also clear that (ii) says exactly the same as Weinberg's condition (1)! Hence, it is hardly surprising that Weinberg's principle of general covariance follows from his equivalence principle (and therefore from Dicke's strong principle), for the former principle actually includes part of the latter!

Misner *et al*'s example shows that a law expressed in tensor form is generally covariant. If we are interested in the validity of these laws we need to ask whether the law is actually *true*. We have seen that the principle of general covariance implies that if the law is valid in *one* coordinate system, it is valid in *all* such systems. But it is equivalence which seems to provide us with the assurance that the laws of STR do apply in local inertial frames, and it is equivalence which tells us that we can set up these frames in the first place. Consequently, the question about the *validity* of the law is dependent upon the truth of the strong principle of equivalence. If this principle is correct in its assertion that local inertial frames can be set up in which the laws of STR hold, then we can go on to say that our generally covariant law of STR seems to be valid. We could, after all, have a law which preserves its form under a general coordinate transformation and which happens to be false.

The question of whether a law is generally covariant is a mathematical matter. If the law is in tensor form then we can answer the question positively. This interpretation follows the standard view, first put forward by Kretschmann (1917), that general covariance is merely a mathematical requirement for a physical theory. Einstein (1918) agrees with this view, which implies that the principle of general covariance is physically empty. Kretschmann observed that we could rewrite any law of physics, expressed in some special coordinate system, in covariant tensor form. Thus, Newtonian mechanics and STR can be rewritten in tensor form. The stricture of general covariance is merely formal, and, in fact, does not even single out GTR as being *the* generally covariant theory. Kretschmann's view has fairly obvious implications for the status of the principle of general covariance in GTR, and we shall consider these later. Here we may note just that the expression of a law in tensor form by itself has no physical ramifications. But the specific numerical content of a tensor for some point in space–time does have physical consequences. Herein lies the difference between equivalence and covariance. A tensor like $T^{\mu\nu}$ in the law of local energy-momentum conservation is not just a mathematical object; it describes a physical state of affairs.† We can

†See Adler *et al* (1975: 117).

test to see if the predicted tensor values in some specified locality hold good. If they do, then we will have empirical support for our assertion that the law is locally valid. This in turn supports our use of the strong principle. We can therefore draw the following distinction between equivalence and covariance: the strong principle is an empirical principle, involving the physical, measurable features of mathematical objects such as tensors; the principle of general covariance goes no further than the insistence that a physical theory of gravitation should be written in covariant tensor form, i.e. the principle is merely concerned with the exploitation of the mathematical properties of the tensor calculus. Weinberg's view of covariance obscures this distinction; if we are to investigate the status of the two principles in GTR without confusion, then we should avoid Weinberg's view of 'equivalence-cum-covariance'. Weinberg succumbs to such confusion himself when he says that *his* principle of general covariance is physically empty. For condition (1) does have physical content, namely the assertion that the laws of STR hold in the absence of gravitation—this is an *empirical* assertion. Consequently, Weinberg can only claim that covariance is physically empty if he drops his 'equivalence' condition (1) from the principle. This would leave him with the mathematical 'covariance' condition (2).

2.2 The status of equivalence and covariance in General Relativity

What of the status of these principles? I shall examine the strong principle of equivalence first. Should we follow those who, like Zahar (1977), see only heuristic value for the strong principle? Of course, there is no doubt that Einstein was guided to the field equations of GTR by the idea of equivalence; but do we have to admit that the strong principle has no place *within* GTR? That is, does the principle express a fact about the gravitational field of GTR, rather than just a fact about how we might arrive at a field theory of gravitation?

In order to settle these questions I shall return to the context of Misner, Thorne and Wheeler's example. If all the non-gravitational equations of GTR were like the law of local energy-momentum in this example, we would have few problems. For it really does look as if the strong principle applies in this case. However, Misner, Thorne and Wheeler point out that their 'commas to semicolons' rule which applies in this example is not a *general* rule to be applied to all the laws of a theory. In the transition from the flat space–time of STR to the curved space–time of GTR, no terms due to curvature appear in the equation. This is not generally the case for all physical laws, e.g. for the transport law for the angular momentum vector $\boldsymbol{S}$ of a spinning

body like the Earth. In flat space–time the transport law is of the form

$$\nabla \cdot \boldsymbol{S} = 0 \qquad (2.10)$$

but in curved space–time a term due to curvature appears in (2.10). The usual source of this trouble for the strong principle lies in the Riemann curvature tensor or its various contractions, e.g. the Ricci tensor. The Riemann tensor $R^{\lambda}{}_{\mu\nu\kappa}$ behaves quite differently from Γ-functions despite the fact that both are derivatives of the metric tensor. The distinction between the R and the Γ is obscured in the flat space–time of STR. In such space–time all geodesics of test bodies appear to be straight: this can only be the case if the Γ disappear from the relativistic equation of motion (2.3). Now, the Riemann tensor is constructed from the $g_{\mu\nu}$ and their first and second derivatives. In flat space–time the components of this tensor are zero, i.e.

$$R^{\lambda}{}_{\mu\nu\kappa} = 0. \qquad (2.11)†$$

In fact, the condition that the Riemann tensor vanish everywhere is necessary and sufficient for the space–time being Euclidean or pseudo-Euclidean, i.e. flat.† Of course, the reason why the Γ and R vanish is that a suitable choice of coordinates in flat space–time makes the g values constant, and therefore their derivatives must be zero. But in the arbitrarily curved space–time of GTR the g are not constant. So, we will expect that the derivatives of the g will not in general be zero. Now, a property of tensors is that if they are zero in any coordinate system they are zero in all such systems; and, if they are non-zero in one, they are non-zero in all. We have observed that the Γ-function can be made to vanish at the origin of a local inertial frame. But because it is *not* a tensor quality, the Γ-function will not in general vanish everywhere in curved space–time. The Riemann tensor must vanish everywhere if it vanishes in any coordinate system; so even in a local inertial frame in arbitrary curved space, the R will in general be non-zero. Indeed the fact that the R are non-zero shows that the metric is curved. Consequently, we should not be surprised when certain laws pick up curvature terms even in local inertial frames. Such terms represent the coupling of the law with the gravitational field. For those laws in which curvature terms appear in the transition from STR to GTR, their precise form is therefore *not preserved*. This is obviously a serious challenge to the status of the strong principle. Unfortunately, there is no firm general rule which we can use to distinguish those laws to which the strong principle cannot apply. Even though curvature terms almost invariably appear in tensor equations which involve derivatives higher than first order, we can only rely upon this as a rule of thumb. Each case has to be

†See Weinberg (1972: 138).

approached on its own merits.† We might try to gain comfort from the fact that for many important laws, e.g. local energy-momentum and Maxwell's first-order equations, no curvature terms need appear. However, if the strong principle is to be accepted as a fundamental principle of GTR, we would expect it to apply to *all* the laws of physics, as it indeed purports to do. It cannot apply in every case, so as it stands it must be incorrect. And, as we have observed, there is no natural way of supplementing the law to resolve this problem, i.e. we cannot say that the strong principle applies to laws which are of such and such a form. If we did try to reformulate the strong principle so that it might thereby have a clear application to some fixed subset of the laws of physics, then the result would be a rather '*post hoc*' principle, involving perhaps quite a number of postscripts to the original principle. This could hardly be called a basic or foundational principle of GTR.

The fact that gravitation, via the curvature terms, cannot be eliminated even at the origin of local inertial frames has prompted Zahar to declare that

> If taken to mean that the (gravitational) field can be transformed away at each point, the principle of equivalence is strictly speaking false. It would, if so interpreted, only apply to flat space. (1980: 29)

So far, Zahar seems to be right. But he goes on to say

> However, the principle of equivalence can be heuristically exploited as follows. We start from a Lorentzian reference system, generate a reducible (gravitational) field by accelerating the frame, study the properties of such a particular field, then go over to the case of an irreducible field. *The last step, however, involves a genuine generalisation.* (1980: 29)

Zahar suggests that the application of the strong principle has the local status of an 'actually false' generalisation. The move to curved space–time is not *strictly* correct. But it is very nearly correct. So, the generalisation in only just off target. Consequently, the strong principle acts as a fairly accurate guide to our target, i.e. the laws of physics in curved space–time. Now, for all I know, Einstein may well have followed something like Zahar's guiding principle. But we can easily avoid the commitment to a false generalisation which Zahar's viewpoint involves and still advocate strong equivalence.

†See Misner *et al* (1973: 391) and also Trautman (1967: 123) and Rindler (1977: 174–7) for a discussion of this problem.

There is another way of seeing the function of the strong principle in GTR. To show this we need only make a slight change in the principle, which may be stated thus

> At each local inertial frame origin, the general definitions and laws of (non-gravitational) physics shall reduce to their special relativistic forms—if possible. (see Rindler (1977: 176))

This form of the strong principle is often referred to as the principle of minimum coupling, for reasons which will soon be evident. The idea of minimum coupling is in accord with Will's (1981) analysis of equivalence mentioned in § 2.1. The function of Will's SEP is to exclude gravitational theories with additional fields whatever their connection with the metric g. The principle of minimum coupling also aims to exclude additional curvature terms: it may therefore be seen as a methodological principle of simplicity. The addendum 'if possible' is all-important. We admit we cannot always perform the reduction to STR laws. But the principle commits us to the attempt to do this. We should therefore only allow curvature terms, which prevent this reduction, to appear in our equations when absolutely necessary.† Consequently, the strong principle acts as a constraint upon us when we are formulating the laws of physics. We should not introduce *ad hoc* curvature terms into our tensor equations, i.e. coupling should be a minimum and only take place when there is a good physical reason. For the addition of a curvature term is an addition of extra physical content. If this content cannot be related exclusively to the physics of the field, then it should be excluded.

If we take the strong principle of equivalence in this 'minimum' sense, then we avoid a commitment to the false generalisation which Zahar's interpretation exhibits. There is no attempt on the part of the principle of minimum coupling to generalise at all. We do not need to worry about the strong principle's 'inexact' application in GTR. For the 'minimum' principle makes allowances for our inability to reach such exactitude. The idea of strong equivalence is used as a *goal*, and the laws of physics should conform to this goal as much as possible. And, we should not forget that we are able to *test* the conformity of our laws to this goal in local experiments.

I offer no comment on the historical issue of which 'heuristic instrument' Einstein really used: it may have been Zahar's principle of equivalence, or the principle of minimum coupling. However, the question of the logical status of the principle of strong equivalence can now be faced. If we interpret the principle as Zahar does, then, as he himself admits, it is not true, and can only be used as a guide to the laws of physics in curved space–time. Zahar also tells us that even

†See Misner *et al* (1973: 390).

used heuristically the principle is false. Its heuristic value comes from the fact that it is *usually* quite accurate. The principle of minimum coupling, however, does apply to all the laws of physics because it is a *methodological* principle. It also incorporates much of the heuristic thinking advocated by Zahar, e.g. studies of flat space–time and of local inertial frames to help us identify the laws of physics. What it does not incorporate is any commitment to the false generalisation recommended by Zahar. I therefore conclude that, if strong equivalence is to have a coherent place in GTR, then it is preferable to use it as the goal of the principle of minimum coupling.

I have already presented the standard view of the principle of general covariance: the demand that the equations of GTR be expressed in covariant tensor form has no physical content and is not unique to GTR. The consequences of covariance being a mathematical requirement, applicable to any physical theory on a differentiable manifold, have obvious implications for the status of the principle in GTR. We can certainly say that the principle is incorporated in GTR; but having said this, we do not seem to have said very much at all. We can hardly maintain that the insistence that equations be expressed in covariant tensor form is a foundational principle of GTR, for there may be other theories which comply with this demand *and* which are inconsistent with GTR.

Because of its physical emptiness and its concern with the laws of physics in general, Bunge has dubbed the principle of general covariance a 'metanomological' principle (1963: 220). Graves, who expresses his agreement with Bunge, says that

> the principle of general covariance is not a particular law of nature on a par with, e.g. the law for the gravitational field. It is, rather, a law of laws, an essentially 'metanomological' principle. Although it is itself a law-like statement, its referents are not objects or events, but physical laws of a lower order. (1971: 194)

Zahar (1980) takes a similar view of the role of the principle; he maintains that the principle is not in the object language of GTR, but in the metalanguage which refers to that theory. Whilst I am in general agreement with them, we should note that the requirement that GTR be expressed in tensor form presupposes the condition that the objects of the theory are to be expressed in geometrical form (see Friedman (1983: 32*f*)). If this were not the case, we would be unable to handle the objects—they *must* be expressed by tensors. So, we must assert that the principle does have *something* to say about the objects of GTR; they must be *geometrical* objects, e.g. the metric, the affine connection, the electromagnetic field and so on. In this sense, at least, the principle of general covariance is not devoid of physical content. For it involves a commitment to a certain kind of physical world: a

geometrical world, based on a continuously differentiable manifold, in which there are no globally preferred coordinate systems. Consequently we must supplement the views of Graves and Zahar with the point that whilst the principle does nor refer directly to the objects of GTR, it nevertheless has direct implications for those objects.

Graves goes on to say that the principle of general covariance 'is the only closure rule necessary for GTR' (1971: 145): so long as we can express its laws in tensor form, we need make no further demand upon GTR. This claim ignores both Einstein's thought on the subject and the final structure of GTR which complies with Einstein's wishes. Einstein stated quite clearly that the laws of physics should be expressed in covariant form and that the equations we arrive at should be '*the simplest ones possible*' (1969: 69). In his paper Einstein listed the implications of this demand for GTR. There are four 'simplifying' requirements which are imposed upon GTR.

(1) The equations of the theory, including the field equations, should be expressed, in covariant tensor form.

(2) Where the Riemann tensor is zero, i.e. in flat space–time, the field equations should admit the solution $g_{\alpha\beta} = \eta_{\alpha\beta}$, i.e. the metric should be that of flat space–time.

(3) No higher than the second derivatives of the metric tensor should be involved in the field equations.

(4) The second derivatives of the metric tensor should enter the field equations linearly to ensure a unique solution.

We have already noted that the substance of requirement (2) is linked to the principle of strong equivalence. This is the only requirement that has an empirical content; when there is no matter, all measurements will reveal that the Riemann tensor has vanished everywhere. Given the adoption of the tensor calculus, requirement (2) takes us very close to the field equations. Einstein drew an analogy between equation (2.11) above ($R^{\lambda}{}_{\mu\nu\kappa} = 0$) and Laplace's equation for matter-free fields in classical physics:

$$\nabla^2\phi = 0 \tag{2.12}$$

where ϕ is the scalar potential whose gradient is the field. In matter-filled space we may use Poisson's equation

$$\nabla^2\phi = 4\pi\kappa\rho \tag{2.13}$$

where ρ denotes the density of matter (see Einstein *et al* (1923: 148)). Hence, requirement (2) leads quite naturally to the claim that the geometry of the field can be related to mass–energy via the field equations, which incorporate, on the one side, tensor quantities

describing the geometry, and, on the other, tensor quantities describing the distribution of mass–energy.

Poisson's equation also provided Einstein with the rationale for requirement (2). Since this equation involves the second derivative of the potential ϕ, it seems quite natural to limit the field equations to second or lower derivatives of the metric tensor. Hence, the Riemann tensor, a second derivative of the $g_{\mu\nu}$, can be expected to enter the field equations.

The first requirement is essentially a restatement of the principle of general covariance, but the other three give substance to the supplementary demand that the equations of GTR be as simple as possible. Whereas (2) is concerned with the content of the equations, (3) and (4) are further mathematical requirements concerned with the form of the equations. Of course, (2) has implications for the final form of the field equations, insofar as they must admit the flat space–time metric $\eta_{\alpha\beta}$ as one of its solutions. In his paper of 1918 Einstein changed his mind about this requirement and altered the field equations with the addition of the cosmological constant, the effect of which was to impose a constant underlying curvature upon the field. Hence (2) may be seen as a formal constraint upon the introduction of such terms, but this has the empirical consequence that we prohibit the introduction of such additional physical content as that which the cosmological constant imports into GTR.†

I hope that this discussion shows that Graves is wrong when he maintains that general covariance is the only closure rule necessary for GTR. The equations of GTR are also constrained by (2), (3) and (4) above. And whereas (2) follows naturally from the strong principle, the essentially mathematical requirements (3) and (4) may be added to the equally mathematical principle of general covariance quite naturally. But talk of limitations on the metric tensor does not limit us to GTR; the conditions (3) and (4) can be applied equally well to STR when stated in covariant tensor form.

Anderson (1964, 1967, 1971) has suggested that if we wish to formulate a principle which satisfies the general postulate of relativity and which is unique to GTR, then we must forget the principle of general covariance as we have stated it, and adopt what I shall call his general principle of relativity (invariance). Anderson's sometimes opaque views are amplified and clarified by Friedman (1973, 1983). These views require the definition of two groups: the covariance group of a physical theory and the invariance group. The covariance group that we have associated with GTR so far is the group of *all* admissible coordinate transformations. But all this amounts to is the demand that GTR's equations be expressed in tensor form. The invariance

†We shall have more to say about the cosmological constant in Chapters 3 and 5.

group is rather more difficult to pin down. Anderson and Friedman try to do this, and show that a principle which relates to this group rather than the covariance group is unique to GTR. Anderson begins with a distinction: between the *absolute* and the *dynamical* geometrical objects postulated by a physical theory. As Friedman explains:

> The absolute objects are thought to be those objects that are not affected by the interactions described by the theory; they are 'independent' of the dynamical objects—part of the fixed 'background framework' within which interaction takes place. (1973: 297)

So, the metric of STR is absolute, but that of GTR is dynamical; time in Newtonian mechanics is absolute, but not in STR. Friedman recommends that we can arrive at the absolute objects of a physical theory by specifying the symmetry group of that theory; the derivation of symmetry groups requires the use of point transformations rather than coordinate transformations. The symmetry group of a theory is defined as the largest subgroup of the group of all differentiable transformations which leaves all the absolute objects of the theory invariant (see Friedman (1983: 56*f*)). These symmetry groups provide us with our 'invariance' groups which list the absolute objects of a theory. Friedman tells us that

> Anderson's program involves construing the traditional relativity principles associated with spacetime theories as requirements on the symmetry groups of these theories. In this interpretation the principle of Galilean relativity says that the symmetry group of Newtonian mechanics is the Galilean group, the special principle of relativity says that the symmetry group of special relativity is the Lorentz group, and the general principle of relativity (invariance) says that the symmetry group of general relativity is the group $\mathcal{M}$ of all differentiable transformations. (1973: 315)

Hence, the three groups maintain the invariance under differentiable transformations of:

(i) for the Galilean group the manifold, the affine connection, time and the metric;
(ii) for the Lorentz group the manifold, the affine connection and the metric; and
(iii) for the group applicable to GTR, only the manifold itself!

The obvious consequence of (iii) is that in GTR there are no absolute objects other than the manifold. Hence, no object other than the manifold is immune from the interactions taking place throughout the manifold.† When we compare the principle of general covariance and Anderson's and Friedman's ideas on symmetry groups and in-

†We shall explore the consequences of Friedman's ideas in § 3.1.

variance, only the latter seem to be unique to GTR. Since the symmetry group of GTR is the group of *all* differentiable transformations, they may seem justified in their claim that this group gives genuine substance to the general postulate of relativity. General covariance is applicable to all physical theories; but according to Friedman the symmetry group above picks out GTR and only GTR.

Anderson's general principle of relativity (invariance) seems to have clear advantages over the principle of general covariance. But I shall maintain that both principles are required in GTR. Anderson's ideas are concerned with the *content* of physical theories: they enable us to judge which objects are absolute in a given theory. But the principle of general covariance is concerned with the *form* of physical theories. It seems perfectly plausible for us to have two field theories both expressed in tensor form and both with exactly the same absolute objects, and therefore the same symmetry groups. However, the form of the field equations differs: one complying with requirement (3) above, and the other involving derivatives higher than the second. Hence Anderson and Friedman's conditions limit the number of absolute objects in a theory, but not the form of the equations of that theory.

Consequently, we must adopt not only the general principle of relativity (invariance), but also the principle of general covariance together with (2) if we are to maintain economy both of absolute content and of form. In their views on covariance and invariance, Anderson and Friedman fail to stress the importance of this requirement. We may therefore conclude that the general postulate of relativity is satisfied in GTR if: the equations are in tensor form; these equations are of the simplest form possible; and the theory involves a minimum of absolute objects.

These demands apply to GTR, and only GTR, *and* they place the required constraints on the form of its equations. They also accommodate Einstein's somewhat notorious confusions about coordinate and point transformations, and about the extent to which covariance and/or invariance is involved in GTR (see Earman and Glymour (1978)). *All* these features are represented in the final form of the general postulate. Moreover, Einstein's repudiation of absolute space understood as an object which influences but cannot be influenced in dynamical interactions is also satisfied.

2.3 Simplicity and relativity

We have seen in the preceding sections a number of mathematical and conceptual devices which help us to arrive at the field equations of GTR.

(1) The idea of weak equivalence which is incorporated in GTR inasmuch as only *one* kind of 'mass' is involved in the theory.

(2) The idea of strong equivalence which is incorporated in GTR in a number of ways: first, inasmuch as only one kind of 'force' is involved in the theory which gains expression in the affine connection ($\Gamma^{\alpha}_{\beta\gamma}$).

(3) Strong equivalence also leads to the use of local inertial frames in GTR which as Misner *et al* point out are simple mathematical devices which allow us to cope with the vagaries of curved space–time.

(4) Again, strong equivalence gains its full expression in GTR when restated as the principle of minimum coupling.

(5) Finally, strong equivalence leads to the demand that terms like the cosmological constant should not enter the field equations, for curvature terms should appear only when absolutely necessary.†

(6) The principle of general covariance is, of course, part of GTR, but when conjoined with the additional requirements constraining the form of the field equations, gives voice to Einstein's demand that the equations of GTR be as simple as possible.

(7) The general principle of relativity (invariance) places a limit on the number of absolute objects in GTR.

We should note a feature of GTR, as it has been expounded so far, which is not clearly Machian. The field equations of GTR may have a preponderance of dynamical objects in them, but this does *not* imply that space–time is relational in a second important sense. We saw in Chapter 1 that Mach wanted to reduce all talk of space (and time) to talk of material objects without remainder. Nothing we have examined, and certainly nothing that Friedman says about the absolute–dynamical distinction seems to satisfy Mach's desire to rid himself of absolute space completely. To do this we would need to show that the metric of GTR is uniquely determined by the mass–energy distribution thus vindicating MP. Whether this is so will be the concern of the next chapter. Nevertheless we are halfway to meeting Mach's demand. The space–time of GTR is not the 'independent' background framework of Newtonian mechanics or STR. And, nothing we have said so far is actually *inconsistent* with Mach's view of space as relational as opposed to absolute in this second sense, and the seven points above are consistent with Mach's hopes for a general theory of dynamics.

However, we should also notice that the various demands of equivalence, covariance and invariance all stress the economical features which GTR should possess. We have every reason to regard the substance of these demands as heuristic principles of economy

†See Misner *et al* (1973: 390) and p.77.

which lead inexorably towards the field equations of GTR. Given Einstein's close acquaintance with Mach's work, it would be indeed surprising if he did not have Mach's theory of economy very much in mind as he advanced towards the field equations. Certainly, we now have every reason to dismiss Holton's implication that economy in science is not a fundamental characteristic of Einstein's thought and work.† Einstein tells us that his progress to the field equations via the ideas of equivalence and covariance is distinguished by considerations of economy:

> I have learned something else from the theory of gravitation: No ever so inclusive collection of empirical facts can ever lead to the setting up of such complicated equations. A theory can be tested by experience, but there is no way from experience to the setting up of a theory. Equations of such complexity as are the equations of the gravitational field can be found only through the discovery of a logically simple mathematical condition which determines the equations completely or (at least) almost completely. Once one has those sufficiently strong formal conditions, one requires only little knowledge of facts for the setting up of a theory; in the case of the equations of gravitation it is the four-dimensionality and the symmetric tensor as expression for the structure of space which, together with the invariance concerning the continuous transformation-group, determine the equations almost completely. (1969: 89)

Hence, for Einstein, as for Mach, economy and simplicity are essential ingredients of successful science. Where they differ, of course, is on the extent to which science should be dominated by observation and experiment.

†See the introduction to this chapter.

Sir Arthur Eddington, a drawing made in 1928 by Sir William Rothenstein. Courtesy National Portrait Gallery, London.

Chapter 3

Mainstream Classical Relativity

Introduction

In this chapter, we shall look at the varieties of classical relativity. The debate between the absolutists and the relationists concerning the nature of GTR space–times will help us to reveal the crucial features of classical relativity. The focal point of this debate concerns the status of Mach's principle in GTR. We should recall from §1.5 that philosophical considerations persuaded us to adopt a version of this principle which does not commit us to the full-blown positivism and sensationalism of either Mach or the later positivists. Rather, Mach's principle should be understood as a methodological injunction, certainly one with a positivistic bias, but which is advocated on the grounds on ontological economy. The job facing the relationist is to show that the empirically plausible solutions of GTR are in line with this injunction, which states:

> *The inertial properties of matter should be uniquely determined by the distribution of matter in the universe, if possible.*

Given the discussion of the last chapter, in which we equated variations in inertial forces with variations in the geometry of space–time, or, more precisely, of the affine structure of space–time, we can restate this version of Mach's principle:

> *The affine structure should be uniquely determined by the distribution of matter in the universe, if possible.*

Note, in both cases, the empirical flavour of the principle: the weak, methodological Machian, or the 'empirical relationist', is concerned

with the physical universe. Hence, if there were to be any substantial empirical evidence for the absolutist counter-claim that the material content of the universe does *not* uniquely determine the affine structure, then the empirical relationist would be forced to embrace the absolute. That is the strength of the rider 'if possible' in the principle. A relationist who advocates Mach's principle on methodological rather than on *a priori* grounds cannot rule out the *possibility* that the essential demand of the principle will not be incorporated in the theory. The relationist case depends on the resolution of several empirical issues: for example, the problem of whether the matter content of our universe exhibits an overall rotation. If the evidence for such a rotation were to be compelling, then there would be little to prevent the advocacy of absolute space–time, in some sense, as a frame of reference for such a rotation.

There are, I think, six questions which are pertinent to this debate between the absolutist and the relationist; examining the possible answers to each will help us to present a coherent survey of classical relativity. We shall also develop a clearer picture of the status of Mach's principle in GTR. And, in keeping with my remarks above, we shall find, in the course of this chapter and the next, that the resolution of each question depends a great deal upon empirical matters. The questions are as follows.

(1) Just how far should we go in imposing casuality conditions and topological constraints upon GTR?

(2) Are there any limitations on the form of the metrics allowable in GTR space–times?

(3) What sort of matter fields can we permit in GTR?

(4) In what form should we express the field equations, which, as we shall see, relate geometrical structure to matter distribution?

(5) Are there any empirical considerations which *force* us to abandon, or alternatively to champion, any of the solutions to the field equations?

(6) What is the status of the cosmological principle in GTR?

In this chapter we shall be concerned primarily with the subject-matter of questions (2)–(6), and in the next chapter with that of question (1). However, the final resolution of these questions will be reserved for Chapter 5.

3.1 The absolute–relational debate and the General Theory of Relativity

The Einstein field equation

$$R_{ik} - \tfrac{1}{2}g_{ik}R = 8\pi T_{ik} \tag{3.1}$$

relates the geometry is space–time, given by the metrical function on the left-hand side of (3.1), to the distribution of matter (or mass–energy), given by the stress-energy tensor T_{ik}. Given that G_{ik}, the Einstein tensor, is specified by

$$G_{ik} = R_{ik} - \tfrac{1}{2}g_{ik}R \tag{3.2}$$

we may rewrite equation (3.1) in its abstract geometric form:

$$\boldsymbol{G} = 8\pi \boldsymbol{T}. \tag{3.3}$$

Henceforth, I shall use $\boldsymbol{G}$ as shorthand for the affine geometrical structure of space–time, i.e. that of the metrical field, and $\boldsymbol{T}$ for the matter distribution. During the course of this chapter we shall employ the concept of a solution of the field equations of GTR, or a *model* of GTR. We shall define a model of GTR as consisting of the following:

(i) a differentiable manifold—the space–time;
(ii) a metric of Lorentz signature on the manifold;
(iii) a matter field which is to be described by the stress–energy tensor; and,
(iv) a set of field equations relating the metric and the matter field.

Hence, a model of GTR is a geometrical model which allows us to describe the dynamics of a given physical situation and to make predictions concerning the dynamical behaviour of physical systems involved in the description. The field equations given above may be regarded as the *general* law of GTR. But when we feed a particular metric and matter field into the general field equations, we obviously limit the scope of those equations to the particular application concerned. We may regard the resultant equations, the equations of the models, as the *derived* laws of GTR. This distinction will become important in the final section of this chapter.

The focal point of the absolutist–relationist debate on the nature of space–time in GTR, is the relationship between $\boldsymbol{G}$ and $\boldsymbol{T}$. In *Space, Time and Spacetime* (1974), Lawrence Sklar adopts a 'substantivalist' or absolutist approach to this problem; this approach is also taken by a number of influential philosophers of physics, e.g. Earman, Gardner and Friedman. It may be conveniently summed up in the maxim: there is more to space–time than talk of material objects. Sklar and the other absolutists give their support to this viewpoint essentially because the two rival programmes—the 'reductionist' or relationist and that of geometrodynamics—seem to have met little or no success. The maxims of these programmes might be, for the relationist: there is no more to space–time than talk of material objects; and for the supporter of geometrodynamics: there is no more to matter than talk

of curved empty space–time. Each position has its own philosophical heritage. Absolutists present Newton as their champion; relationists prefer Leibniz or Mach; and the geometrodynamicists opt for Descartes.

The claim that space, or space–time, is absolute is, however, confusing. There are numerous senses of 'absolute' to be found in the literature on space and time. In *The Measure of the Universe* (1965), North has unearthed eight or nine senses of 'absolute' in various contexts, and he admits that his list is far from exhaustive. As if to prove this point, Earman, in 'Who's afraid of absolute space?' (1970), provides us with a list of a dozen senses. And they show that although the space of one theory may be shown to be absolute in a particular sense, the space of another may not be absolute in that sense. Both North and Earman warn us that we need to know the sense of absolute we employ, and the theoretical context with which we are concerned. We just can't expect answers to vague questions like 'is "space" absolute?', unless we specify the theory with which we are concerned and which way we are using 'absolute' in our question. Moreover, we can't expect historical debates like that between Leibniz and Clarke to be entirely relevant to a discussion of general relativistic space–times. Nor should we think that Mach's sceptical remarks about 'absolute' space apply without modification to GTR.

A useful analysis of the ideas of absolutism and relationism is provided by Newton-Smith (1980). There have also been several other comprehensive analyses of the various meanings of absolute and relational, particularly those by North (1965), Earman (1970), and Hooker (1971). Although Newton-Smith's work is directed at the nature of time, it can easily be extended to cover space and space–time. He contrasts Platonism with reductionism. Essentially the Platonist is an absolutist who grants ontological independence to time and maintains the 'absolute' necessity of time's topological properties as *a priori* constructions. The reductionist, or relationist, resists this strategy, arguing that time depends on things and the relations between things, and that topological assertions about time are contingent rather than necessary truths.† This contrast allows us to outline standard views taken by absolutists and relationists. The absolutist may believe that:

(A1) There is a four-dimensional arena—space–time—which consists of spatio-temporal points existing independently of objects in space–time.

(A2) The topological properties of space–time are independent of objects in space–time.

†See Newton-Smith (1980: 9–10).

(A3) The metrical properties of space–time are independent of objects in space–time.

(A4) The spatio-temporal relations between bodies depend only on the spatio-temporal relations between points in space–time.

The reductionist or relationist may believe that:

(R1) There is a four-dimensional system of objects in space–time.

(R2) The topological properties of space–time depend on the objects in space–time.

(R3) The metrical properties of space–time depend on the objects in space–time.

(R4) Space–time is nothing more than the system of objects: hence all spatio-temporal relations are reducible to relations between objects.

Although committed absolutists and relationists might try to adopt all four of their respective beliefs, both camps will find it hard to carry through their programmes. Of course, if an absolutist were to be committed to Newtonian gravity, then (A1), (A2), (A3) and (A4) might be said to stand. But, as soon as we move into the context of non-Euclidean and pseudo-Euclidean gravitational fields, such simplicity is lost. For the metric is now a dynamic object which is absolute in some senses and relational in others. Consequently, we require an analysis of absolutism and relationism which highlights the differences between the various elements of modern gravitational theories.

I believe the most appropriate account is that given by Michael Friedman in his *Foundations of Spacetime Theories* (1983). Friedman sets down three central distinctions:

(i) absolute$_1$–relational;
(ii) absolute$_2$–relative; and
(iii) absolute$_3$–dynamic.

It is the first distinction which provides fuel for the debate which is our concern. The other two give rise to little or no controversy in the context of GTR: the second is not directly relevant to the debate, and although the third sheds light on the debate, it gives comfort to both parties.

3.1(a) The absolute$_1$–relational distinction

When we ask whether space–time is absolute$_1$ or not, we wish to know whether spatio-temporal properties and relations are ontologically reducible to materially based properties and relations without any remainder. That is, if we believe that such a reduction is possible, then we might, like Reichenbach (1958), take our basic properties to

be causal, or like Mach (1960) take them to be phenomenal. These are relational or reductionist viewpoints. Their most perplexing problem is the need for an explanation of inertial forces in terms of matter distribution. They cannot rely on the simple idea of relative acceleration, for this does not tell us why the passenger in the train leaving the station experiences inertial forces, but the people on the platform do not: the relative acceleration is the same for passenger and those on the platform. The usual path chosen by the relationist is to argue that inertial forces are determined by the motion of bodies and systems with respect to the entire matter distribution in the universe. Now, in GTR we explain the physical properties of motion in terms of the geometrical (affine) properties of curves on the differentiable manifold. In one respect this is an antirelationist move: for motion and its properties are explained geometrically via the affine connection. A trajectory which is not a geodesic of the connection on the manifold represents the non-inertial motion of a test particle; the geodesical trajectories represent the inertial motions. We may regard the affine connection as a field which determines the paths of test particles at all points of space–time, i.e. as the metrical field.† However, the relationist can argue that these metrical characteristics are determined by the matter distribution above.

There seem to be two kinds of absolutist$_1$ positions: strong and weak. The strong absolutist$_1$ maintains that no reduction is possible for a given theory of space and time. For example, with regard to his own theory, Newton was a strong absolutist$_1$: he saw no way in which absolute rotations could be reduced to rotations relative to a material frame of reference. Space for him had an unassailable position. Nobody today seems quite so willing to take up such a strong position with respect to GTR. The weak absolutist$_1$ usually believes that a reduction may be *possible* in the context of a given theory, but he sees no reason to abandon his absolutist$_1$ predilections: the onus is on the relationist to show that such a reduction can be effected. Consequently, most weak absolutists$_1$ seem to base their absolutism upon the failure of the relationist/reductionist programmes. In the case of GTR, this attitude is evident in a large number of recent studies, e.g. in Gardner (1977), Grunbaum (1973), Sklar (1974), Friedman (1983). We shall consider in §3.2 why they believe these programmes to have failed.

3.1(b) The absolutist$_2$–relative distinction

When one talks of space and time being 'relative' notions in STR and GTR, it is in the sense that they are not well defined independently of

†The affine function is a derivative of the metric tensor. See §2.2

frame of reference: for example, the notions of sameness of spatial position, and of sameness of temporal position, can only be defined in a frame-dependent way in the above theories. In Newton's theory, however, they are well defined independently of reference frame: here at least they are absolute$_2$ elements of the spatio-temporal structure. In relativistic space–time theories we can only define the above notions relative to a given frame of reference. So in GTR these notions are not absolute$_2$ elements of the theory, they are relative elements. But there are absolute$_2$ elements in GTR, e.g. the sameness of *space–time* position, which is frame-independent.

3.1(c) The absolute$_3$–dynamic distinction

Newton says that absolute rotation is an empirical fact: we can detect such rotation via our observations of inertial forces. Newton believed that space is 'absolute' because it is independent of the motions and changes of objects within it: it has a fixed and unchanging structure. The combination of these two beliefs seems to imply that Newtonian space can act upon certain bodies and systems—producing inertial effects—but that space itself suffers no reciprocal effects. Newton also held that a reaction must accompany every physical action, and he expressed this in his third law of motion. However, we can see that despite the action of space upon rotating systems, there is in this case no reaction proposed by Newton. These considerations introduce us to the third distinction. An element of the geometrical structure is absolute$_3$ if it is fixed independently of the material events and processes which occur within space–time and the elements of the geometrical structure not so fixed. Consequently, the metric (Euclidean) of Newton's space is absolute$_3$ since it is independent of the material contents of space: it is the metric which is fixed and unchanging. But in GTR the metric is 'affected' by the matter distribution, and it is therefore dynamic. This dynamic nature of the metric is shown clearly by the Einstein field equation: $\boldsymbol{G}$ depends upon $\boldsymbol{T}$, and vice versa. The virtue of GTR for an absolutist is that it removes any discomfort that might be felt about a space which acts but is not acted upon. The space–time of GTR is far from Newton's independent and fixed arena. In fact, the only element of the geometrical structure of GTR which remains absolute$_3$ is the differentiable manifold itself, and of course the geometrical characteristics of space–time *qua* manifold, e.g. topological characteristics such as orientability, completeness and so on.†

†Sklar provides us with a useful review of these characteristics, together with affine characteristics (1974: 46–54). We shall examine the ideas involved in some of these topological notions in Chapter 4.

Friedman goes on to show that these three distinctions are independent of each other. I shall not repeat his demonstrations which are fairly straightforward (see Friedman (1983: 64*f*)). But we may observe that there is nothing we have said about GTR in respect of the absolute$_2$–relative and the absolute$_3$–dynamic distinctions which might offend the absolutist$_1$ or the relationist. The job facing the relationist is to show that the affine properties of space–time can be uniquely determined by the matter distribution, i.e. to show that we need no more than a specification of the matter distribution to say all there is to say about the space–time metric. But although there is a mutual dependence of matter and geometry in GTR, i.e. the metric is *not* absolute$_3$, this by itself does not guarantee that there is a unique determination of $\boldsymbol{G}$ by $\boldsymbol{T}$. It is this problem which we will examine in §3.2.

The account which I have presented does not really cover the reductionism proposed by the geometrodynamicist. Given the tremendous difficulties this programme has experienced, I shall not try to fit it into my view of the absolute–relational debate (but see Stachel (1972) *The Rise and Fall of Geometrodynamics*). This leaves us with two main positions in the debate, one relational and one (weak) absolutist.

(1) The 'Machian' or empirical *relationist*: essentially this relationist believes that $\boldsymbol{T}$ and $\boldsymbol{G}$ are functionally related via the field equations and he aims to reduce $\boldsymbol{G}$ to $\boldsymbol{T}$.

(2) The 'Sklar' *absolutist*$_1$: like Sklar, he believes that the field equations express a relationship of lawlike co-existence between $\boldsymbol{T}$ and $\boldsymbol{G}$; his 'dualistic' beliefs persuade him to resist the idea of a reduction of $\boldsymbol{G}$ to $\boldsymbol{T}$.

3.2 The problems facing the followers of Mach

Although GTR shows us that the structure of space–time is related to the distribution of matter, the Einstein field equations place few restrictions upon the kinds of space–time and the kinds of matter which might be related by the equations. Hawking and Ellis point out that *any* space–time metric could be regarded as satisfying the field equations. They warn that

> . . . having determined the left-hand side of [the field equations] from the metric tensor of the spacetime (g_{ab}), one can *define* T_{ab} as the right-hand side. . . . The matter tensor so defined will in general have unreasonable physical properties; the solution will only be reasonable if the matter content is reasonable. (1973: 117)

Physicists and mathematicians have therefore focused their attention upon the so-called 'exact' solutions of the field equations. We postulate that the space–time of GTR solutions is a four-dimensional Riemannian space $(\mathcal{M}, g)$ with metric normal form: diag. $(1,1,1,-1)$. For exact solutions, the field equations are satisfied with $\boldsymbol{T}$, the stress-energy tensor of some specified form of matter, which obeys two basic conditions; a local causality condition and an energy condition. *Local causality* demands that the equations governing the matter field must be such that the curve along which a signal travels between two points is a non-space-like curve (see Hawking and Ellis (1973: 59*f*)). *The energy condition* requires that the energy density in a given region should not be exceeded by the pressure in that region, i.e. that kinetic energy in a locality should never be negative (see Hawking and Ellis (1973: 88*f*)). Hawking and Ellis say that these conditions are physically reasonable inasmuch as they hold for all known forms of matter. It is our wider (than GTR) physical knowledge which indicates that this is the case. Hence, the conditions are physical restraints which we impose upon GTR to ensure at least a minimum compatibility between GTR and our other accepted physical theories: we restrict the space–time metrics to those which do not lead to marked conflicts with the accepted view of matter and its behaviour.† However, these are fairly minimal constraints and a wide range of GTR exact solutions are available to us. The objections to the relationist or reductionist views are usually based on a liberal attitude to what counts as an 'admissible' solution of the field equations. The two conditions cited above require only that the matter content of GTR space–times obeys certain rules, but not that there should actually be a matter content. The absolutists certainly make use of such empty space–times in their arguments. In fact, there are three types of solution which are used as ammunition against the relationist:

(*a*) 'empty' space–time solutions;
(*b*) mass solutions in which the matter content does not uniquely determine the metric structure; and
(*c*) rotating mass solutions.

As we shall see, each of these kinds of exact solution involves space–time essentially and irreducibly and each therefore challenges the relationist assertion that the matter distribution is all we need to determine all motions in the space–time.

One of the most comprehensive attacks made on the relationist position we have described above is made by Sklar in *Space, Time and Spacetime*. We shall consider how he and other absolutists use the three

†We might also note that the field equations themselves require that $\boldsymbol{\nabla}\cdot\boldsymbol{T} = 0$, i.e. that energy-momentum is locally conserved.

types of solution (*a*), (*b*) and (*c*) given above to highlight the weaknesses of the relationist.

3.2(*a*) *Empty solutions*

Newton's absolutist views seem to require that Newtonian space has a structure which is independent of matter and which provides us with standards of absolute zeros of both translational and rotational accelerations. Even with no matter content this structure would still obtain, and the concept of acceleration would be well defined 'absolutely'. We have seen that the mutual dependence of the metric and the matter distribution in GTR seems to offer some comfort to the relationist. For there now appears to be a distinct possibility of showing that we only need the matter distribution to define standards of inertial and non-inertial motion. This would be in line with the idea, expressed in Mach's principle, that the explanation of inertial forces should not require an appeal to any irreducible element of space–time. However, most physicists now believe that Mach's dreams are not fully realised in GTR. For example, Hawking and Israel tell us in the introduction to *General Relativity: An Einstein Centenary Survey*, that

> Einstein was strongly influenced by Mach, and believed that general relativity incorporates his Principle. However, this is not the case, because the Einstein equations admit many solutions, including the flat Minkowski metric, which contain no matter at all. In fact, by making the metric a dynamical field, Einstein removed the distinction between matter and the metric, which embodies the structure of spacetime. (1979: 12)

Sklar also takes this position. He maintains that the relationist's belief that inertial forces are caused by *relative* motion (to the overall matter distribution) has the consequence that in an empty space–time there could not be two distinct states for test systems, namely inertial and non-inertial. The relationist would expect all test motions to be inertial, if anything. But Sklar says that since the empty 'Minkowski' solutions of the field equations are 'perfectly admissible' in GTR, a privileged frame of reference is determined for motions (see Sklar (1974: 216)). In the flat Minkowski solution the motions of test particles are everywhere determined; and this seems to be clearly non-relational, for what could determine the motions other than the geometrical structure of space–time? The trajectories of particles, in this flat space–time, which are not geodesics are those which we regard as experiencing inertial forces—the particles are moving non-inertially. These forces and motions are now interpreted geometrically via the affine connection, which defines the 'straightest' curve

between two points of space–time. Those particles which follow the geodesics of the connection are those which we regard as moving inertially, and free of all external and centrifugal-type forces. So, in this 'empty' Minkowskian solution, both inertial and non-inertial motions are perfectly well defined. The Machian claim that we should deal only with motion defined relative to material systems seems inappropriate with respect to this empty space–time.

The usual response to such arguments is to deny that empty solutions are realistic. Bondi does this in *Assumption and Myth in Physical Theory* (1967). His reasons are based on the belief that GTR is primarily an empirical theory grounded in observation and experiment. He says, with clear echoes of Mach,

> We expect a decent theory to go a *little further* than the experiments and observations on which it is based . . .. But we never expect our theories to hold in circumstances utterly and completely different . . .. To expect that a theory so based should work in the absence of a material universe seems to me an entirely unjustified step. (1967: 75–6)

Mach, of course, in his criticism of Newton's views on space, time and motion maintained that we could not entertain the possibility of empty or virtually empty universes. We should face the fact that the universe is given, 'fixed' stars and all, empirically.

Mach's attitude receives a strong rebuff from Bertrand Russell in *The Principles of Mathematics* (1903). Since Russell's opinions are essentially those of Sklar and other modern absolutists, I shall set them down here. Mach claims in *The Science of Mechanics* (1960) that we cannot draw any physical inference from Newton's thought experiments. Newton, in the second of these experiments, had asked us to imagine two globes joined by a cord. If we observe the system in the laboratory (which is assumed to be at rest or in uniform motion relative to the 'fixed' stars) then we would find that there are two possible states of affairs: when the globes are rotating about a common centre we will observe a tension in the cord; when there is no rotation relative to the laboratory, there will be no tension. Newton thought that it is quite natural to regard the presence of force in the cord as a sign of rotation. In the laboratory there is a material frame of reference to which we can refer the motion of the system. However, if we imagine the experiment taking place in an otherwise empty universe, and if tension is observed in the cord, the only inference Newton believes we can make is that the system is rotating with respect to a non-material frame of reference, i.e. the motion is absolute. In this way Newton tried to persuade us that absolute motion is not just a metaphysical notion but has measurable physical effects, i.e. inertial forces. There is, of course, no difficulty, for the

absolutist at least, in utilising Newton's 'two globes' test system in the flat Minkowski space–time of GTR. If we wish to protest at this move, we will need to show that the idea of an empty space–time is physically meaningless, which is, of course, the issue at stake.

Russell reminds us of Mach's response to Newton's argument:

> [Mach] remarks that, in the actual world, the earth rotates relative to the fixed stars, and that the universe is not given twice over in different shapes, but only once, and as we find it. Hence any argument that the rotation of the earth (or any other system) could be inferred *if* there were no heavenly bodies is futile. (1903: 492)

Russell thinks that Mach's preoccupation with the actual world is dangerous, and says that it seems to be at odds with scientific practice. Mach's claim is that since the 'fixed stars' are actual existents we cannot ignore this fact when we try to account for the appearance of inertial forces in rotating systems. But scientists invariably concern themselves with the possible as well as the actual. Russell says that scientific laws

> .. can be applied to universes which do not exist . . . it is evident that the laws are so applied [in dynamics], and that, in all exact calculations the distribution of matter which is assumed is not that of the actual world. It seems impossible to deny any significance to such calculations . . . this being so, the universe is given, as an entity, not only twice, but as many times as there are possible distributions of matter, and Mach's argument falls to the ground. (1903: 493)

Russell is asserting, just as many absolutists before and after him, the right to (i) utilise any model consistent with the laws of a dynamical theory without regard to the particular initial and boundary conditions of the model and (ii) to draw physical inferences from models so used. The second right is essential to the absolutist position. We shall see that a relationist could grant the first, but deny that inferences drawn from 'unrealistic' models have any *physical* significance. The relationist might argue that such models have only a *pragmatic* value, merely helping us to clarify our ideas about the 'realistic' models. We shall return to this argument in later sections. Mach, however, seems to deny (i): according to his account it seems that we are only entitled to consider the model universe which describes the actual, observed physical universe. However, Mach is far from being clear about what is and is not of physical significance. Nevertheless it is possible to detect four distinct kinds of physical possibility in *The Science of Mechanics*.

(1) Those states of affairs which may be logically possible, but have no physical significance, direct or indirect, e.g. that described in Newton's 'two globes' experiment.

(2) Those which are 'idealisations' which help us to elucidate the actual indirectly, despite the fact that they are not to be found in the actual world, e.g. those described by the ideal gas laws.

(3) Those situations which are 'realistic' physical possibilities, but which have not been, are not and may never be actual. Such situations allow us to make direct inferences about the actual world by virtue of their consistency with both scientific 'laws' and, broadly speaking, the initial and boundary conditions of the physical world as discovered by observation and experiment. That is such possibilities challenge neither the 'laws' nor the observed facts (e.g. most of the calculations we make when theorising about what would have occurred, or what would occur in certain specified circumstances).

(4) Those states of affairs which are actual. That is the observed facts (e.g. the relative positions of the fixed stars and the observed motions with respect to them).

So, contrary to Russell's beliefs, Mach is not altogether rigid in his views about what counts as physically significant. Like Bondi, he allows physics to go a *little further* than actual observation and experiment. But how do we know when we have crossed the divide between the realistic and the unrealistic, or between the significant and the physically meaningless? Our only guides seem to be the observed initial and boundary conditions, and the accepted physical laws. If the world were a static 'once and for all' given universe, then we might have some hope of formulating a precise criterion of physical significance. The Machian might say that a basis for such a criterion is the observed, tremendous matter content. To deny that this enormous bulk distribution exists would be to deny also, by fiat, the Machian's claim that the inertial forces observed in rotating systems arise because of the relative rotation of those systems with respect to the distant, scattered stars. However, we must note a concession which is urged upon the Machian position by the claims of modern cosmology. Even though the Machian might maintain that the observed *amount* of matter must be taken into consideration in our physical models, we must recognise that the precise 'scattering' of matter is not so important. Most cosmologists would now accept that the observational evidence points towards an 'evolving' universe involving, perhaps, dramatic changes in the distribution and character of matter, as the universe expands from the initial singularity (see, for example, Sciama's (1971) *Modern Cosmology*). It is now quite natural for cosmologists to talk about the different ways in which the

universe might evolve after a 'big bang', using model universes, like the Friedmann models, to illustrate their various scenarios. The range of alternatives before us seems to be indicative not only of the mathematical ingenuity of cosmologists, but also of a certain amount of insecurity they feel about the observational foundations of the models. But, of all the models of GTR, it seems reasonable to suppose that the cosmological models are perhaps more 'realistic' than certain other model universes. This is an issue to which we shall have to return.

The Machian might demand that our models must incorporate the observed amount of matter, or something very close to this, even if he is not certain what distributions must obtain in permissible models. We can envisage a cosmologist obliging him in this respect, but still causing him displeasure. The cosmologist might construct a model universe in which the 'observed' bulk matter is divided equally between two enormous spheres, at a very great distance from each other, which, for some inexplicable reason, are connected by a cord. But this seems to be the Newtonian 'two globes' universe! We can even imagine the two spheres possessing scientific communities, one of which might be 'Newtonian' and the other might know of GTR. Both sets of scientists discover a tension in the cord. Both reason that the tension demonstrates that their 'universe' could be rotating, i.e. that the matter content is exhibiting an overall rotation. For the Newtonians, the 'two globes' picture as described by Newton is adequate to this situation. For the relativists, a 'two globes' test system in an otherwise flat Minkowski space–time seems adequate to describe the dynamics of their universe. And, as we have observed above, such a space–time seems to be non-relational in character. But, in this relativistic case, the model universe seems to have at least some cosmological, and therefore physical, significance. The Machian might protest that we are using a 'test system' to represent the situation; but this is a fairly common ploy in dynamics, and certainly one which Mach does not object to in *The Science of Mechanics*. The Machian's only recourse seems to be a further entrenchment in the 'realm of observation': he must say that the physically admissible models are those which have as high a degree of consistency as is possible with the observed initial and boundary conditions. Cosmologists would probably agree that the 'two globes' model is not really a candidate for a description of the actual universe at any stage of its evolution. But we shall see that even the 'mass' solutions of GTR cause the Machian some embarrassment.

3.2(b) 'Mass' solutions without unique determination of the metric: the problem of boundary conditions

If inertia is to depend uniquely on the matter distribution, as relationists must insist, then the metric must be uniquely determined

by the matter distribution. If we specify some ***T***, then one and only one ***G*** must be determined by the field equations. Sklar tells us that, in general, the ***T*** function does *not* uniquely determine the metric function ***G***: the reason for this lies in the fact that we must make a choice of our boundary conditions. This means that Mach's principle must not be seen as consisting of the demand that ***T*** uniquely determines ***G***; for this is clearly not the case. Rather we should say that *for a given space–time* (including boundary conditions) ***T*** must determine ***G*** uniquely. Consequently, the choice of boundary condition becomes important for the relationist. Why does such a choice need to be made, and how do we make it? Mathematically, at least, the answers are fairly straightforward. In his 'Mach's principle as boundary equations' (1964), Wheeler tells us

> Einstein's equations are not enough. Differential equations do not define a solution by themselves. They must be supplemented by suitable boundary conditions. (1964: 306)

We have already observed that cosmologists feel themselves free to choose from a wide range of initial and boundary conditions for their models. Considered as a mathematical problem, this choice gives them no particular worries. But, in general, cosmologists wish their models to be empirically adequate. So the boundary conditions chosen are usually those which are candidates for those which actually obtain in the universe. What they are, however, is not so easy to determine: so the choice remains fairly wide. Not only are the initial conditions of our universe something of a mystery—our knowledge of the first hundred seconds after the initial singularity is far from adequate—but also there is no clear consensus concerning the actual boundary conditions. The boundary conditions may be 'Minkowskian' at infinity, or the universe may be spatially closed (see, for example, Rindler (1977: 238*f*), and Misner *et al* (1973: 549*f*, 1118)).

Sklar maintains that there is, however, one particularly natural way for us to view the field of GTR and the boundary conditions we impose upon it. For Sklar 'the natural boundary conditions' specify that 'spacetime is Minkowskian in regions far removed from any matter immersed in it'; this means that

> The behaviour of a test system in a region affected by the nearby presence of matter will then depend upon (1) the fact that spacetime is 'overall' Minkowskian and (2) the fact that the presence of mass-energy has 'distorted' the local spacetime from its natural Minkowskian flatness. (1974: 219)

But we know from our discussion above that in Minkowskian space–time both inertial and non-inertial motions are well defined

without reference to material bodies essentially because the space–time has a global inertial frame of reference which is non-material. Sklar tells us that

> . . . the distortion from flatness of a region of spacetime due to the imposition of gravitational effects on the system over and above the normal *inertial* effects of the pervasive Minkowski spacetime. (1974: 219)

Consequently, test systems experience both gravitational forces, from the nearby matter, and inertial forces because of their acceleration relative to the inertial field defined by the Minkowski space–time. Sklar admits that on his view GTR differs from an STR plus 'gravitation' account only in the *details* of the forces predicted. There is no qualitative difference. Sklar goes on to defend this view:

> The fact that the full specification of the spacetime 'field' in general relativity can be determined only after boundary conditions are imposed, and the fact that it is this element that leads to most of the non-Machian aspects of the theory, may give the non-Machian features the appearance of more mathematical technicalities. I think this is false. In classical field theories there is a 'natural' field, the zero-field, and the role of charges on sources of fields is to superadd their own contributions over and above the 'ground level' zero-field. In general relativity the natural source-free field to adopt is that of Minkowski spacetime, a spacetime in which absolute accelerations are, as we have seen, both well defined and empirically detectable. It is this physical feature of gravitation being superimposed upon an already present inertial field, that the necessity for the impositions of boundary conditions mathematically represents. (1974: 220)

I shall try to put Sklar's picture of GTR in context, for he does not admit how contentious his views are, inasmuch as they follow closely the fairly controversial opinions of Weinberg and other particle theorists.

In Chapter 2, we observed that the Lorentz metric tensor, g_{ik} of GTR takes on the values of the Minkowski metric tensor, η_{ik}, of STR *locally*. The affine function Γ^i_{jk} vanishes at this locality, but because it is not a tensor quantity it will not in general vanish everywhere in curved space–time. The Γ-function is a derivative of the metric function, and since the g-functions will not be constant in the arbitrarily curved Riemannian space–time of GTR, their derivatives will not in general be zero. The Γ-function will only vanish everywhere in space–time if $g_{ik} = \eta_{ik}$ *globally*. However, we know from observations that in the actual universe we cannot assume global frames (see the details of the

Pound–Rebka redshift experiment in Misner, Thorne and Wheeler (1973: 1056–60)). Indeed, the arbitrarily curved space–time of GTR would lead us to expect this. As a consequence of this fact, one might feel that it is natural for us to drop the idea of a 'global' Minkowski metric from GTR. However, Weinberg and other particle theorists refuse to do this (see, for example, Weinberg *Gravitation and Cosmology* (1972)). Rather than accept g_{ik} as *the* metric of GTR, they take g_{ik} to be a tensor field 'painted' on to the Minkowski space–time of STR. That is, instead of regarding the Minkowski metric as *one* possible solution of GTR, they regard it as a metric which underlies all solutions. They do this because of the advantages this approach has for their work in particle physics and electromagnetism: although STR is unable to provide an adequate account of gravitation by itself, since light 'bending' round the sun does not move along the geodesics of η_{ik}, it does give a good account as far as high-speed particles are concerned. And, as Sklar points out, the idea of a source-free field works well in electromagnetism. We can now see why Weinberg *et al* might be eager to retain the idea of an underlying η_{ik} field.

We can easily detect the origins of Sklar's picture of GTR. Like Weinberg, Sklar accepts the g_{ik} field as one that is painted or superimposed upon the η_{ik} field. Consequently, the removal of matter from space–time does not lend to the disappearance of the space–time structure: the inertial field defined by η_{ik} will be left. This gives the Minkowski metric an essential part to play in GTR. However, although the space–time of special relativity is an example of a Riemannian space–time

> . . . it is rather a special example, since it is not only Riemannian, but globally pseudo-Euclidean. (Weinberg 1972: 113)

But Weinberg and Sklar single out η_{ik} and give it a special ontological status in GTR. This picture conflicts with that given by Hawking and Israel. Whereas Weinberg and Sklar offer us two metrics, the η_{ik} and the g_{ik} as well as the matter field given by the stress-energy tensor $\boldsymbol{T}$, Hawking and Israel believe that GTR eradicates the differences between matter and metric. For this latter pair, the metric g_{ik} embodies the structure of space–time. The metric η_{ik}, seen as an underlying structure, seems to be an unnecessary addition to GTR; for g_{ik} incorporates the idea of an *arbitrarily* curved space–time, of which η_{ik} is just one special case. The best that can be said of the η_{ik} field is that it provides an instrumental link between GTR and other field theories. However, without any precise unifying links between these different physical theories, it is perhaps something of an overstatement to say that the *natural* view of GTR is that of Weinberg and Sklar.

Rather, the natural view is that of the g_{ik} field, and this has the distinct advantage of ontological economy: we are no longer committed to the *existence* of an underlying η_{ik} field. This does not deny the physical significance of solutions where $g_{ik} = \eta_{ik}$ globally, or where space–time takes on the 'flat' Minkowski values at infinity. The question of whether such solutions have physical significance is yet to be resolved, but each is a candidate for the appellation of 'physically significant'. What I am resisting is the adoption of an underlying source-free field in GTR, a field which 'underwrites' all of our solutions. The metric which underwrites the solutions of GTR is that defined by g_{ik}. No other metrical concept is needed.†

Sklar also implies that the 'natural' boundary conditions for GTR are the Minkowski values at infinity. This follows from his arguments above concerning the metric and his statement that 'the natural boundary conditions' are that 'spacetime is Minkowskian in regions far removed from any matter immersed in it'. If he were right then the 'natural' solutions of GTR would be non-relational in character. The properties of motion would be determined by reference to the resultant 'overall' Minkowski structure of space–time, and not to the matter distribution alone. But we could opt, like Wheeler (1964), for spatial closure as our 'natural' boundary condition. This brings us back to the problem with which we started. Knowledge of matter distribution is not enough; if we are to determine the properties of space–time and motion we must also impose boundary conditions. The obvious relationist answer to this problem is that our observations of the actual universe should determine the conditions appropriate for GTR models. Mach wished to use physics in order to discover the facts about physical reality: the actual boundary conditions are fairly important facts, for their choice affects the metrical structure of space–time. Although $\boldsymbol{T}$ does not always uniquely determine $\boldsymbol{G}$, $\boldsymbol{T}$ plus the boundary conditions do. The relationist requirement is that we use the observed boundary conditions. However, as we observed in the discussion of cosmological models above, we are uncertain about the actual conditions. Sklar's strongest argument against relationism is that there are solutions of GTR with a matter content, but with 'flat' space–time at infinity. We shall consider later whether any empirical evidence could possibly dissuade us from holding Sklar's 'natural' viewpoint. Certainly, if we allow such solutions to stand, then they prevent any straightforward relationist account of GTR. The response demanded from the relationist is three-fold: the relationist must provide a set of boundary conditions which are not antirelational in character, but which support his relationist views; these conditions

†Sklar has told me that despite appearances to the contrary he did not mean to imply that the η field has anything more than instrumental value.

must have a firm empirical basis; and fairly clear reasons must be given for rejecting other antirelational solutions.

3.2(c) Rotating solutions

Sklar tells us that for the relationist an absolutely rotating 'universe' should be an impossibility. By this, Sklar must mean that there should be no overall rotation of the matter field in a 'relationist' model universe. For with respect to what is the rotation taking place? Space–time structure is the only answer. But Sklar maintains that there exist solutions of GTR in which the matter content exhibits an overall rotation; he gives the Gödel model as an example. The peculiar properties of this model, however, are often cited as a reason for dismissing it. Hawking and Ellis say that the closed time-like curves of the model imply that 'the solution is not very physical' (1973: 170).† But there are other solutions which exhibit overall rotations, e.g. Kerr's solution and, on one view, the Oszvàth and Schücking solution. The initial relationist response to such solutions may be to ask whether the idea of the 'universe' rotating is empirically plausible. Now the Friedmann models of cosmology assume that there is *no* overall rotation, and this assumption, just as that of expansion of the universe, is based on empirical evidence. Recent studies show that the margin of error for the assumption of no rotation of the overall matter content of the universe (i.e. for the assumption of isotropy) is $\leqslant 0.1\%$ (see Collins and Hawking (1973), Sciama (1971: 30) and Will (1981)). The possibility of such a rotation seems negligible. However, this fact questions only the cosmological significance we might give to rotating solutions. Hawking and Ellis remind is that 'In general, astronomical bodies are rotating' (1973: 161). Therefore, we should not be too hasty in ruling out *all* rotating solutions on empirical grounds. The predictions from such solutions as Kerr's metric may be perfectly adequate as a physical description of the behaviour of certain isolated, rotating bodies. Indeed, such solutions seem to be the only ones which can explain the properties of singularities of gravitational collapse (black holes) (see Hawking and Ellis (1973: 161*f*)). Again, the onus is on the relationist to tell us why such solutions are not physically admissible.

In our remarks on general mass solutions we saw that the relationist is going to have to deliver a physically acceptable set of boundary conditions which do not show inconsistencies with his relationist beliefs. Einstein once believed that the natural boundary conditions of GTR are those which specify spatial closure. This idea

†North goes further. He says that such curves show the solution is 'absurd' (1965: 306). We shall return to this question in Chapter 4.

has been taken up by many, including Wheeler, and the initial value formulation (IVF) of GTR gives us perhaps the most acceptable models incorporating the closure conditions. Wheeler had proposed that we should regard Mach's principle as providing

> . . . a boundary condition to separate allowable solutions of Einstein's equation from physically inadmissible solutions (1964: 306)

And the demand for spatial closure is considered by many relationists to be consistent with Mach's principle and therefore to provide the required relationist criterion of admissible solutions of GTR. These solutions would presumably be those consistent with the IVF approach, in which space–time is divided into spatially closed hypersurfaces. The difficulties in doing this should not be underestimated, and I shall consider some important problems for the IVF approach in §3.3 following. But, for the moment, we will grant that the difficulties facing the IVF solutions are merely technical and can be solved in principle.

A spatially closed universe seems to be an attractive candidate for a Machian model universe: there is no question of varying the global properties of space–time without some observed effects upon the matter content (see Misner *et al* (1973: 1181*f*)). Therefore, only one space–time metric structure could be consistent with the field equations and a given matter distribution: there would be a unique determination of ***G*** by ***T***. Sklar remains fairly neutral on this issue; he accepts that spatial closure may pay dividends for the relationist, but he warns that

> . . . it is still far from clear that the resulting modified theory is yet in conformity with Machian expectations in their fullest extent. (1974: 221)

One such expectation is that Machian models should not allow overall rotations of the matter content.

In the *Foundations of Spacetime Theories* (1983), Michael Friedman argues that the condition of spatial closure does not vindicate relationist hopes. He shows that the Oszvàth and Schücking solution exhibits a rotation of the matter field (or matter contents) despite the fact that the model is spatially closed and consistent with the IVF approach (see Oszvàth and Schücking (1969) and Friedman (1983: 210)).

There seem to be two possible strategies of defence for the relationist.

(1) The relationist could argue that we should supplement the

spatial closure condition with a requirement that the matter field should not rotate in admissible solutions. Although this appears to be somewhat '*post hoc*', the empirical support for this additional condition is greater than that for the condition of spatial closure. As we have observed above, we are not certain that space is closed, but we are reasonably confident that the material contents of the universe are not in overall rotation. Indeed, this is one of the 'twin' assumptions, made by cosmologists, of isotropy and of homogeneity, which form the basis of the so-called cosmological principle. This principle demands that all galaxies stand in the same relation to the whole universe (see Rindler (1977: 15, 202)). The Oszvàth and Schücking model is homogeneous, with dust-filled space-like hypersurfaces, but is non-isotropic. The problem for the relationist, however, is that he is now going to find it difficult to describe the dynamics of isolated rotating bodies. By ruling out 'rotating' solutions, he has left himself without any clear-cut explanation of such astronomical phenomena. Again, the onus is on the relationist to provide a quantitatively accurate account of rotating systems within the limits set down by his essentially cosmological approach to solutions. Hawking and Israel (1979: 13) tell us that cosmological models may be 'successful in describing the large-scale structure of the universe' but that they are not easily modified to 'reflect the local irregularities, such as stars and galaxies' which do, in general, rotate.

(2) The relationist could make more specific attacks in the Oszvàth and Schücking model, and try to rule it out on general physical grounds. The most obvious feature of this solution which would give us grounds for attack is its use of the cosmological constant Λ. We might note that Sklar's example of a 'rotating' solution, Gödel's model, also employs Λ. Rather than issue a blanket prohibition against rotating solutions, we might deny instead that models which incorporate Λ lack physical significance. As we pointed out in Chapter 2, Λ represents a source-free constant curvature which pervades space–time. The constant is only used in matter-filled solutions or in 'negative pressure' vacuum solutions.† In the former class of solutions, part of the curvature is due to matter and part is due to Λ. However, Hawking and Israel tell us that

> The motivation for the cosmological term (has) disappeared, and it is now generally ignored. Measurements of distant galaxies place an upper limit of 10^{-66} cm^{-2} on $|\Lambda|$. (1979: 13)

Given this empirical backing, we could simply stipulate that Λ should

†For example in Schwarzschild and Reissner–Nordstrom solutions modified to include $\Lambda \neq 0$, Λ represents the negative pressure.

be set at zero in the field equations. This straightforward action rules out the embarrassment caused by the Gödel and Oszvàth and Schücking models. But Gardner, in 'Relationism and relativity' (1977), says that

> . . . this Machian reply turns out to be insufficient . . . given a solution with $\Lambda = 0$, one can regard it as a solution with $\Lambda \neq 0$ by the simple expedient of transporting to the other side of the field equation and absorbing it into the stress-energy tensor. (1977: 224)

Gardner admits that resulting stress-energy tensor $\boldsymbol{T}$ may not be very plausible physically, but he implies that physical plausibility is not a good enough criterion for the admissibility of GTR solutions.

It is not only relationists who will find this attitude odd. But what happens when the Λ term is on the right-hand side of the field equations? If we represent the matter field of such a solution by $\boldsymbol{T}$, and then remove all 'measurable' matter, i.e. that which we usually associate with the stress-energy tensor $\boldsymbol{T}$, then we are left with a matter field of a sort, $\boldsymbol{T}(\Lambda)$. However, Gardner does not tell us why we should suppose this excess 'matter' to exist. There seems to be no observational reason for us to think this. Nor are there any clear-cut theoretical advantages. Indeed, there may be definite disadvantages, given that we might have to explain away matter with negative energy density, which will result from the transportation of some Λ quantities into $\boldsymbol{T}$. So, whichever side of the field equation Λ enters into, the consequences are unsatisfactory.

Each of these strategies depends upon the familiar relationist reliance on the security of observation and experiment. In his repudiation of global anisotropy and Λ the relationist seems to be on fairly safe ground. However, we have noted that any general prohibition made against overall matter field rotation has the unfortunate consequence that the relationist finds it difficult to account for non-cosmological rotations in general. Of course, this difficulty may only be a technical matter which the relationist can resolve. But the onus is upon him to do this, if he can.

Perhaps the weakest part of the IVF relationist account is the reliance on the condition that the universe be spatially closed. To deny the possibility of openness is a grave failing for what is allegedly an empirical approach to GTR especially when recent work suggests that the universe is spatially open (see Gott (1976) and Penrose (1979)).

At the end of §3.2(b) we made three demands of the relationist. We can see that, so far, he has not given a good account of himself.

(1) We asked him to provide a set of boundary conditions which support his relationist views: the condition of spatial closure was

found to be inadequate, and the relationist has to supplement this condition with the requirements of global isotropy and of $\Lambda = 0$. In his favour, we discovered that these requirements have strong empirical backing.

(2) We demanded that the boundary conditions have a firm empirical foundation: despite the fact that the idea of spatial closure is not empirically otiose, the relationist fails to take account of the strong likelihood that the universe is spatially open.

(3) We asked him to give fairly clear reasons for his rejection of 'absolute' solutions: this is a tremendously difficult enterprise. There are quite a few non-relational solutions which do seem to have some astronomical or local application, e.g. Kerr's metric or the Schwarzschild solution. The predictions made from such models seem to be perfectly respectable from an empirical point of view. We observed that the relationist account tends to rely on cosmological evidence to support its position. But, for example, the Schwarzschild solution models the geometry of the solar system to a very good approximation (see Hawking and Ellis (1973: 143)). The cosmological models are not always good indicators of local behaviour. A lot of work would have to be done to modify IVF-type approaches so that local irregularities could be accounted for. Only then would we feel justified in dispensing with some very useful solutions.

There are two additional minor concerns for the relationist.

(1) The stress-energy tensor $\boldsymbol{T}$ itself contains a metric function; the relationist will have to specify how we can 'separate' this metric function from the 'material' aspect of the tensor if we are to have a purely matter-based quantity. This may, however, be simply a technical problem.

(2) In practice we never arrive at exact solutions by specifying $\boldsymbol{T}$ and the boundary conditions, and then determining the metric. Rindler tells us that in general we have to work on both sides of the field equations at the same time:

> . . . the symmetry of the physical solution may suggest a certain pattern of g's involving unknown functions; the T's can then be expressed in terms of these functions, which are finally determined by the field equations when the T's and g's are substituted into them. (1977: 183)

Of course, this is only a practical problem for the relationist, and has no implications for the possibility of a reduction of $\boldsymbol{G}$ to $\boldsymbol{T}$.

The absolutist seems to have few worries about the boundary conditions we might impose on the field equations. Sklar calls upon the Minkowski solution, the Gödel model and matter-filled solutions which are flat at infinity to support his antirelationist views. Indeed,

the absolutist could object to the relationist programme which seeks to restrict admissible solutions to those with particular initial and boundary conditions. But this objection has little force, for even the absolutist imposes constraints upon GTR. Apart from metric conditions, in GTR conditions are imposed on the matter field and on the topology. Hawking and Ellis, although sceptical about the place of Mach's principle in GTR, begin their discussion of the solutions of the theory by specifying the two matter field conditions cited at the beginning of this section. Moreover, the topological conditions of geodesic completeness, orientability and so on are usually imposed upon space–time. If we did not impose these two kinds of constraint upon GTR, tremendous theoretical complications would arise. And these complications would have *no* empirical foundation: we have absolutely no observational reason to drop the above conditions. So, if we can impose essentially empirical conditions on the matter field and on the topology, why should the relationist not urge us to choose empirically based boundary conditions? But given the observational evidence, the choice would have to be more flexible than that made in the IVF approach. In §3.3 I shall consider Raine's approach to the problem: not only does this have the finest empirical pedigree, it enables us to account for most cosmological variations.

3.3 Cosmology and relativity

In his article 'Mach's principle in General Relativity' (1975), Raine employs Mach's principle as a criterion for selecting admissible cosmological solutions of the Einstein field equations.† As is usual, he demands that the metric in admissible solutions arises from material sources alone. Despite the exceptional technicality of the paper, Raine's conclusions have tremendous empirical plausibility. Raine presents a modified field equation, based on the work of Sciama *et al* (1969). He shows that we can specify the metric of a space–time from the integral of the matter source terms and Green function. Normally we can only use such an integral method in a linear theory, and GTR is non-linear. But despite the non-linearity of the modified field equations, they possess linear characteristics which facilitate integration. Raine also gives us reassurance by showing that the Einstein field equations can easily be recovered from a limiting value of the modified equations.

Raine imposes two conditions of uniqueness upon the solutions of the modified field equations. The first is concerned with the

†Further comments on this work are given in Raine (1981).

relationship of the metric and the curvature. His 'Machian' prejudice is that physicists observe curvature, not the metric, which should be derived from the curvature. Raine then demands that a specified curvature should determine a unique metric for the solution to be admissible. This condition is not satisfied when the curvature is zero, so we can rule out solutions with zero curvature on mathematical grounds. Raine shows that empty space–times and asymptotically flat space–times fail to satisfy this first condition. We saw in §3.2 above that such solutions are essentially antirelational; being able to rule them out in a straightforward way is certainly an advantage for the relationist. The second condition is concerned with the relationship between the matter distribution and the curvature. Raine demands that given a particular matter distribution the Weyl curvature tensor should be uniquely determined. Between them the Ricci and the Weyl tensors determine the curvature of space–time. We can determine the Ricci curvature from the field equations, but the Weyl curvature is not so obviously given. Hence the need to determine the Weyl tensor uniquely from the matter distribution. If we cannot show such a unique determination, then once again the solution may be ruled out. As with the first condition, this condition also enables us to rule out empty space–times.

The two conditions together ensure a unique determination of the metric by the matter distribution (via curvature) in admissible solutions. Raine shows that there are two main kinds of admissible solution: Robertson–Walker models and Bondi spherically symmetric solutions. Both kinds are 'Machian' or relationist, in the sense that in them the matter distribution uniquely determines the metric. Raine argues that spatially homogeneous models with rotation or anisotropic expansion are ruled out by the above conditions.

3.3(a) Robertson–Walker models: the Friedmann solutions

If we assume that space is homogeneous and isotropic, then considerations of symmetry enable us to determine the structure of space–time uniquely given 'the density and motion pattern at one cosmic instant' (Rindler (1977: 243)). It is therefore not surprising that the Friedmann models (which have a Robertson–Walker metric) satisfy Raine's Machian conditions. These spatially homogeneous and isotropic models have been very successful in describing the large-scale structure of the universe, and, as Hawking and Israel tell us,

> the Friedmann models are a good approximation at least back to the first one hundred seconds. (1979: 4)

Friedmann's solutions allowed for an expanding universe with a

uniform distribution of matter, from an initial singularity. Friedmann's work, published in 1922,

> ... fitted the observations of the time, and indeed all subsequent measurements of the large scale structure of the universe. (Hawking and Israel (1979: 13))

The solutions also allow for the possibilities of open and closed universes. As we have observed above, which 'universe' we are in in still in dispute. We know the present 'rate' of expansion $\dot{l}_0$ given by the Hubble constant, but this leaves the question open (see figure 3.1). We don't have enough information to determine the shape of the curve: (A) represents an undecelerated expansion or an open universe; (B) represents a recontracting or closed universe. Nevertheless, most cosmologists would concur with Hawking and Israel that the Friedmann models provide our best basis for the description of the universe as a whole. However, their allegiance to the cosmological principle, which demands a high degree of symmetry in the models, prevents them from giving any straightforward account of local irregularities.

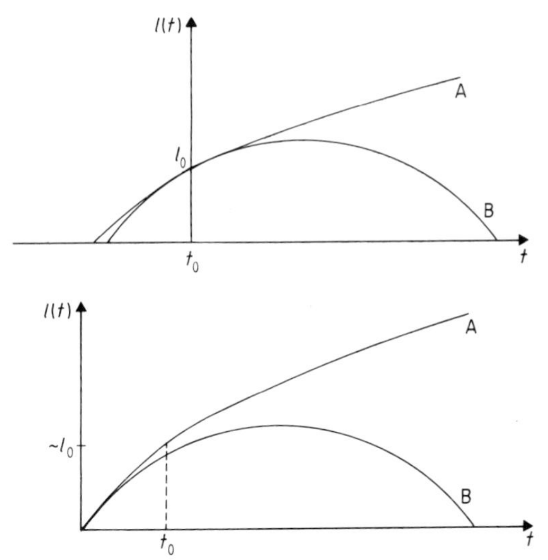

Figure 3.1 The scale factor $l(t)$ of the universe for an open model (A) and a closed model (B). If we assume that $\dot{l}_0$ is the same for A and for B, then the upper graph applies with uncertainty over the age of the universe. But if we admit a degree of uncertainty for $\dot{l}_0$, then we can produce a common origin for both A and B, as in the lower graph. In fact, there is some uncertainty over the present rate of expansion, and so this must result in some uncertainty over the age of the two models.

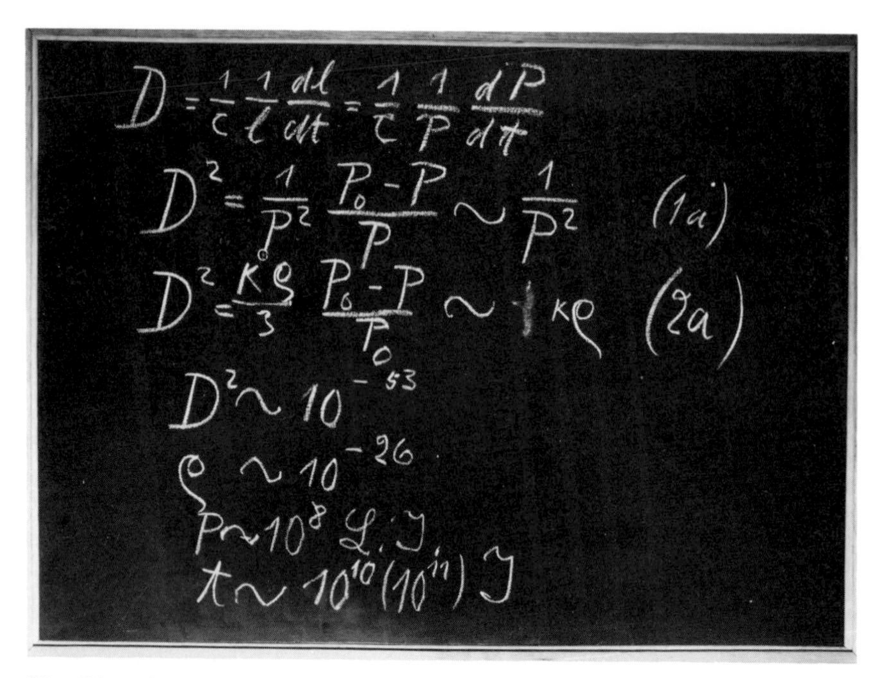

The Einstein blackboard. The Einstein lectures at Oxford were given at a time of great excitement in the astronomical world. The blackboard preserves, in Einstein's hand, some of the ground covered in his lectures in 1930. The topic of this particular lecture was the expansion of the universe. The ground covered in this lecture is described by Einstein and De Sitter in a joint paper 'On the relation between the expansion and the mean density of the universe' *Proc. Natl. Acad. Sci.* **18** 213–14 (1932), and this is also the topic of this lecture. Courtesy Museum of the History of Science, Oxford.

3.3(b) Bondi spherically symmetric solutions

This failing on the part of the Friedmann models may be partially remedied by the Bondi models which are *not* spatially homogeneous. Again, the symmetry involved in these solutions leads us to suspect that they will satisfy Raine's Machian conditions, as indeed they do (see, for a discussion of spherical symmetry, Rindler on the Schwarzschild solution (1977: 136*f*)). Raine says of the Bondi models, which also exhibit an expansion from a (peculiar kind of) initial singularity:

> While they are not realistic models of the actual universe, they show interesting features which may be expected to be present in more complex situations. (1975: 521)

Since the actual universe is locally inhomogeneous, the Bondi models may prove useful for a Machian account of these local distortions from the overall homogeneity of space.

The absolutist$_1$ has three main lines of defence.

(1) The first is based on the empirical inadequacies of the Friedmann models. Raine offers us no clear-cut way of describing the dynamics of local irregularities: the offer of Bondi-type models to assist in some such situation does not guarantee explanations of all local phenomena. Of course, Raine says he is primarily interested in showing that the large-scale features of the universe can be described by a Machian theory. But such a theory ought to be able to give a clear account of local, as well as global, dynamics.

(2) The second is based on a theoretical problem. What should we say about the absolutist$_1$ models and solutions which are ruled out by Raine's Machian conditions? Although they are insufficient to ensure unique determination using Raine's modified equations, their results still seem to stand for unmodified GTR. If these models served no useful purpose we might be content to accept Raine's verdict and rule them out. But given that Raine cannot account for all local dynamical situations, the absolutist$_1$ models are needed. For such local situations, we will, in general, have to use unmodified GTR: the onus is on the relationist to show that we can dispense with absolutist$_1$ solutions without theoretical loss, and Raine does not do this.

(3) The third 'philosophical' problem for the relationist hits at the heart of Raine's account. We are happy with the results he presents primarily because they are in line with present day cosmological theory. But how secure is this theory? How much credence should we give to the central ideas of overall homogeneity, isotropy and expansion? In his article 'In defence of induction' (1980) Mackie maintains that the assumption of a stable, simple universe is rational, for we have little or no reason to suppose that the universe is chaotic. However, it is one thing for use to assume that the universe is not chaotic—on the evidence we have, the probability if a chaotic universe is near zero—but it is quite another for us to maintain that the universe has a particular kind of simplicity, and we are doing just this when we proffer the Friedmann solutions as models of our universe. The Friedmann models, with their assumption of the cosmological principle, are the modern realisation of the Copernican belief that the human race is not at the centre of the cosmos: homogeneity and isotropy demand that no place in the universe should have a privileged status. But this assumption raises two questions as follows.

(1) Does the cosmological principle apply at *all* times during the expansion of the universe? We know that the Friedmann models give

us good approximations back to the first 100 seconds, but did the universe exhibit the characteristics of homogeneity and isotropy at earlier times; and will it do so at very late times? That is, could not our current observations of the universe merely be giving us access to a privileged time range in the history of the universe, namely those times at which it appears very nearly homogeneous and isotropic? Indeed Sciama (1971) and Zel'dovich (1979) maintain that at such times the universe must be irregular.

(2) Does the cosmological principle apply at all? Ellis in his 'Is the Universe, expanding?' (1978) maintains that we can drop the principle, build an inhomogeneous *static* (non-expanding) model, and still account for all the observations which cosmologists cite in favour of the particular kind of simplicity involved in the Friedmann models. Although Ellis does not ask us to accept his model as being that of the actual universe, he does ask us to be a little more critical in our acceptance of the assumption of homogeneity.† So, the absolutist$_1$ could argue that the success of Raine's account depends upon an assumption which might involve a considerable over-statement of the homogeneous character of the universe.

In answer to these criticisms the relationist must fall back on the status of the Friedmann solutions in GTR as the best cosmological models we have. Despite their inadequacies as far as local irregularities are concerned, no other group of GTR models offers us a satisfactory account of cosmological observations at all but early times. The Ellis model, whilst it purports to do this, must be seen as a tentative model of the universe, given that its empirical and theoretical implications have not been fully explored. The Friedmann models, however, have stood the test of over sixty years. Nevertheless, the absolutist$_1$ objections should leave us with some unease at the prospect of putting our *complete* faith in these models. Perhaps the most powerful of these objections seems to be the fact that 'absolutist$_1$' solutions of GTR do have applications, despite their dismissal by Raine's criteria.

3.4 The laws of relativity

We have observed two main weaknesses in the case for empirical relationism: absolutist$_1$ solutions do seem to have an application in dynamics—even if we try to dismiss them on mathematical grounds; and Newton's two globes thought experiment is a powerful weapon for the absolutist$_1$.

These two problems for the relationist turn on the same issue: can

†A description of the Ellis model is given in the Appendix.

we impose limits upon the admissible solutions of dynamical theories? Raine suggests that we can do so, and he aims to show that mathematical criteria allow us to select the admissible models of GTR. Like Wheeler and many others who have argued for relationism, Raine believes that MP may be used as a criterion of model-choice in GTR. We have seen that many such attempts have not been altogether successful, e.g. the IVF approach mentioned in §3.2(c). But there remains the possibility that GTR will never accommodate the demands of the relationist. If the relationist is not able to demonstrate that 'empty' space–time and other problem solutions *cannot* be used in GTR, then he is in a dilemma. Either he must abandon his commitment to MP and embrace absolute space–time, or he must search for a new, acceptable theory which conforms to MP. If the relationist decides to hang on to the MP, then he will be using the principle as a criterion of theory choice. And he would then have to offer some very good reasons to justify his rejection of GTR. It would hardly be enough for him to say simply that GTR is not in accord with MP, especially if the theory retains the respect and attention of the scientific community. We noted in Chapter 1 that the empirical relationist is entitled to his deep suspicion of 'theoretical' entities, but this does not empower him to rule out a theory which is satisfactory in all respects other than its use of the concept of absolute space or space–time. The relationist would need to spotlight problems and deficiencies in GTR on a wider front; and he would then have to show how the resolution of such difficulties would be facilitated by the formulation of a new, *relationist* theory. Given that few would be bold (or foolish) enough to reject GTR in the present empirical and theoretical climate, the relationist might be better advised to remain faithful to GTR. If he does so, then the need to rule out problem solutions is acute.

At the heart of the problem lies Newton's two globes experiment. As we have discovered, in §§1.3 and 3.2(a), Newton's argument gives a philosophical respectability to those solutions which perplex the relationist. This respectability, together with the fact that these absolutist$_1$ solutions do have their uses, challenges the relationist who seeks to rule out absolutist$_1$ models of GTR. So, if the relationist is to have a chance of winning the debate, he must undermine the respectability which Newton gives to absolutist$_1$ solutions.

We should first recall the conclusion of §3.2(a): absolutist$_1$ solutions are consistent with the field equations of GTR—in particular, there seems to be no cogent reason for the repudiation of a solution in which two globes in an otherwise empty space–time exhibit an overall rotation. All the Machian, or empirical relationist, can do is protest that the absolutist$_1$ solutions do not have a sufficiently strong empirical foundation, that they lie outside the empirically acceptable range of possibilities in dynamics. But, as yet, this protest has not

been given a sharp enough edge to enable us to remove the offending solutions from the context of GTR. Although Raine might seem to come close to vindicating the relationist case, he fails because he cannot demonstrate that absolutist$_1$ solutions have no application in GTR whatsoever; so long as they have an application—and so long as Raine's methods do not provide us with acceptable, relationist alternatives—we cannot make the move from the orthodox approach to GTR to Raine's integral methods.

Accordingly, if we are to undermine the respectability of the absolutist$_1$ solutions, we must show that they do not have applications in dynamics. The purpose of this section is to provide some substance to the relationist's contention that they have no empirical foundations. I shall not confine myself to the argument that the 'evidence' is against such solutions. For we have seen that the evidence, although it favours the relationist, does not do so unequivocally. Rather I shall argue that the field equations of GTR must be constrained by conceptual as well as empirical considerations, and that our attention should be directed towards the derived laws of GTR, and the physical situations they describe, rather than towards the more general field equations.

There is in Newton's two globes experiment the presumption that our dynamical laws have an unrestricted range of application. We have seen that Mach resists this lack of constraint. He demands that we take into account the actual configuration of matter in the universe. In response to this demand, Russell argues that we can utilise *any* model consistent with the laws of a theory and that we may then draw physical inferences from these models. It is this argument which must be challenged.

A standard formulation for a general scientific or natural law is:

$$(x)(Fx \rightarrow Gx).$$

The law as stated—with no riders, implicit or explicit—seems to commit us to taking all its consequences seriously, even when these lack empirical plausibility. But we have seen Hawking and Ellis impose restrictions on a general law, demanding that only certain kinds of matter field should be employed in GTR.† Why should we not impose other restrictions? What are the rules or guidelines which we should follow when seeking to impose such restrictions? Can we provide a distinction between acceptable and unacceptable restrictions on the scope of a general law? And can we make the distinction clear enough to help us arrive at a straightforward resolution of the absolute–relational debate?

Consider a space–time with a region with negative kinetic energy.

†See Hawking and Ellis (1973: 117) and §3.2.

What laws obtain in the space–time? We discover that the field equations of GTR are adequate to describe the physics of the space–time. So why do we rule out solutions with local negative kinetic energy? Two reasons may be advanced: no one has observed negative kinetic energy; and there doesn't seem to be any convincing wider theoretical justification for negative kinetic energy which brings us no benefits in GTR, and which doesn't fit in with our physical ideas in general. Hence, ruling out negative kinetic energy is consistent with our broad empirical view of the physical world.

Consider a theory about water (H_2O). On the basis of the theory we ascribe various properties to water in various states. A law of the theory might tell us how the behaviour of water varies with temperature. So, we might say:

(1) If the temperature of this water were 300 K then its properties would be such and such.

But the law and theory seem to allow us to say:

(2) If the temperature of this water were 10^{50} K then its behaviour would be such and such.

But what sense is there in talking about temperatures of 10^{50} K—leaving aside the question of whether we can talk of 'water' at such temperatures. For even at times immediately after the initial singularity the temperature may not exceed 10^{32} K (see Gibbons, Hawking and Siklos (1985)). There seems to be a point, albeit 'fuzzy', at which the law becomes inapplicable. Hence, the law seems to have a *range* of applicability. Indeed, no one could seriously propose that laws involving temperature apply at negative absolute temperatures. Just as no one could seriously talk of temperatures *far* in excess of 10^{32} K.

Such limitations seem to be implicit in all laws. That is, any law has an implicit range of application; and no one would seriously propose a law without some implicit notion of its range of applicability. For any law, we could find some extreme conditions in which it would break down or be quite inappropriate. But this does not mean that the proposer of the law is going to think that he has made a major mistake in advancing the law. Just because 'black holes' are classified as 'places' where the laws of GTR break down, we do not abandon those laws. The physicist realises that the law is not intended to obtain in such conditions. The advocates of the law never intended that the law should be asserted in such extreme conditions.

Who or what decides the range of application of a law? It looks as if this is done by the scientific community with an eye on the wider

theoretical context in which the law is meant to apply. Because the theoretical context of laws involving temperature gives no physical meaning to negative absolute temperatures, physicists have no reason to assert that any law obtains at such temperatures. Similarly, since cosmologists see no justification for the claim that temperatures much more than 10^{32} K obtain even just after the 'hot big bang', there seems to be no reason to assert that any law obtains at temperatures in excess of 10^{32} K. Of course, cosmologists might discover that their account of the early universe needs radical revision, and they might decide that the initial temperatures are of the order of, say, 10^{35} K. But in such a case, the theoretical context would be changed—and this change would permit the extension of the range of applicability of the relevant laws. Laws are integral parts of theories and changes in the theoretical context may have dramatic consequences for the law and its use. If we discovered that we could not describe the dynamics of the universe without the cosmological constant, then we would make an *explicit* change in the field equations. But if we discovered that negative kinetic energy is empirically plausible, we would make an *implicit* change in the range of application of the field equations—just as Hawking and Ellis constrain the range of the laws of GTR by imposing implicit limits on the matter fields permissible in GTR space–times.

Hence, a change in the theoretical context may result in an appropriate change in the form and implication of a given law and/or a shift in the use of the law by the scientific community, with a corresponding change in the implicit range of application of the law.

The change in the theoretical context may be brought about not only by empirical findings but also because of relatively high-level theoretical considerations. This is because the theoretical context itself is a complex organism which includes specific empirical and theoretical statements, as well as general laws. Here, I reveal my commitment to the Quinean 'network' view of science (see Quine (1953) and §5.4).

Consider the debate concerning the origin of the universe. Some physicists are still resistant to the idea of a universe expanding from an initial singularity (see, for example, Hoyle (1975) and Narlikar (1978) who have advocated a model of the universe involving the continuous creations of mass-energy with no initial singularity). But most cosmologists accept that the discovery of the 'microwave background' is evidence for a universe evolving from a radiation 'soup' just after the 'big bang' to the present matter-dominated era (see Sciama (1971) and Raychaudhuri (1979)). When the scientific community accepted the now orthodox interpretation of the micro-

wave background and its place in the history of the universe, a limit was thereby imposed on the theoretical context; and consequently our laws of gravitation became constrained by the demand that cosmological models should take account of the known properties of the microwave background.

Therefore, the fact that laws have an implicit range of application provides a rationale for the scientific community to set aside certain models. Sometimes the charge is made that the dismissal of various models by the relationist is an *ad hoc* move. But the theoretical context gives the relationist a *conceptual* structure which enables him to rule out those models in definite conflict with that structure as unsatisfactory and empirically unacceptable. Hence, it is not just a matter of appealing to empirical facts in isolation to justify the repudiation of a given absolutist$_1$ solution. The relationist appeals to the overall conceptual structure—of which empirical evidence is a part—if he wishes to support the claim that a particular model is unacceptable. And an intrinsic part of the conceptual structure is the range of application of each law contained in the structure. Accordingly, the decision to rule out a model may be made because the physics of the model conflicts with the conceptual demands of the theory concerned, and therefore because the model lies outside of the range of application of the laws of the theory. This is quite clearly what happens in the case of models of GTR with locally negative kinetic energy.

Of course, we may not always be able to take such a clear-cut decision. I have already admitted that a range of application will be 'fuzzy' at its edges. This is because our empirical knowledge and theoretical understanding is not unequivocal. There is room for doubt and disagreement. But the relationist will be inclined to insist that debate should be limited to the problems about which there is a reasonable doubt. This allows him to exclude all physical possibilities which conflict *decisively* with the theoretical context; such possibilities, whilst they may retain a definite physical meaning,† may be regarded as physically implausible and outside the scope of the laws of the theory. I must therefore conclude that Russell's claim, that we can utilise and draw inferences from *any* model consistent with the laws of a theory, carries little conviction. For this claim ignores the fact that a law is part of a theoretical context, which constrains the range of application of the law. Therefore, not only can we justify the actions of physicists like Hawking and Ellis who seek to exclude certain matter fields from GTR, but we can also give support to the general relationist

†They will have physical meaning in the sense that we can imagine a physical theory with the same laws but set in a different theoretical context and with a wider range of application.

strategy of attempting to rule out absolutist$_1$ solutions. But we should note that support for the strategy does not commit us to support for the repudiation of any given model. The onus is on the relationist to show that the various absolutist$_1$ solutions should be ruled out. And perhaps the crucial task will be to face up to the first weakness mentioned at the beginning of the section: the fact that certain absolutist$_1$ solutions seem to have an application in GTR.

So, what should the relationist say about those absolutist$_1$ solutions which do seem to have an application, namely those which are concerned with local irregularities? To resolve this problem, we need to look more closely at the theoretical context of GTR, and, in particular, examine the role of the Robertson–Walker solutions. Consider the reasonable assumption that the metric of the actual universe as a whole is very nearly Robertson–Walker. What do we then make of the supposition that the metric is *not* Robertson–Walker or anything like it?

Our answer must depend on the condition of the theoretical context. If there are numerous worries concerning the status of Robertson–Walker models, or if alternative models have an increasing amount of empirical and theoretical support, then we would be inclined to say that the supposition has a degree of plausibility. But in these circumstances the theoretical context would be in a state of some disarray. The consequence of such confused circumstances would be that the range of application of the field equations would be widened. For example, if we are no longer convinced that locally negative kinetic energy has little empirical plausibility, then we cannot rule out applications of the laws to models with negative kinetic energy. But we have seen that as far as the Robertson–Walker metric is concerned, the theoretical context seems to be unambiguous: solutions with this metric have a pre-eminent status. If we are to describe the observed global mass–energy distribution in terms of GTR, then we need the Robertson–Walker metric—or something very close to it.

When we plug the Robertson–Walker metric into the field equations of GTR, we arrive at derived laws which play a crucial role in the global description of the universe. As far as most cosmologists seem to be concerned, the models with Robertson–Walker metrics, particularly the Friedmann models, are the centre of attention—perhaps more than the Einstein field equations themselves. They have confidence in Robertson–Walker solutions, and the theoretical context of gravitation and cosmology underpins that confidence. Indeed, we have seen that Raine is willing to tinker with the Einstein field equations, and he offers as partial justification for his modifications the fact that the modified equations deliver the foremost solutions of modern cosmology. He is ready to change the general

laws of the theory, and claims vindication when the change leads to the well entrenched, derived 'laws of cosmology', i.e. to the equations which provide the basis of the Friedmann models.

It is hardly surprising that physicists like Raine who work in cosmology have a good deal of sympathy for relationism. Their attention is directed towards the Friedmann models, the metrics of these models are Robertson–Walker, and Robertson–Walker space–times are relationist solutions. Physicists who are inclined to oppose absolutist$_1$ solutions, such as the Kerr model, might well believe that such models do not fit in with the global cosmological context. They might find it hard to make good sense of rotating mass solutions when the matter in our universe does not exhibit an overall rotation. Or they might be sceptical about empty space–times given that the universe is patently occupied. But their real complaint cannot be simply that such solutions are *used* to describe local gravitational phenomena. It must be the case that they distrust the inference from a local solution which neglects the cosmological context to an assertion about the global nature of space–time.

The absolutist$_1$ argument seems to run as follows.

(1) Some GTR solutions are absolute$_1$.

(2) There is a reasonable amount of empirical and theoretical backing for these solutions—they describe local phenomena quite adequately.

(3) Therefore, we need to use absolute$_1$ space–times in GTR.

(4) Despite the fact that *some* solutions do not require an irreducible reference to absolute$_1$ space–time, GTR as a whole requires absolute$_1$ space–time as an irreducible element in our description of gravitational phenomena.

The relationist counteracts this argument by challenging step (4). The relationist's counter-argument runs as follows.

(1) An implicit assumption of the relationist position is that the global cosmological context must be taken into account in all solutions.

(2) Local solutions tend to neglect this context.

(3) Global solutions, like the Friedmann models, do not nelgect the cosmological context.

(4) Appeal to the cosmological context by itself does not enable us to rule out absolutist$_1$ solutions, for they are obviously useful.

(5) But appeal to the cosmological context does permit us to block the inference from the *use* of absolutist$_1$ models to the claim that absoiute space–time is an *essential* ingredient of GTR.

(6) An implicit assumption of the absolutist$_1$ position seems to be that the cosmological context need not be taken into account—his

solutions may stand in isolation and carry the same physical force as relationist solutions.

(7) But this assumption is unwarranted, since when dealing with the gravitational field we must do so in a global context. Of course, we can use a model to describe the behaviour of a single rotating star; but we should not forget that such phenomena are integral parts of the wider gravitational system.

(8) Because absolutist$_1$ models neglect the cosmological context, they must be treated as at best mere approximations.

(9) So, properly speaking, absolutist$_1$ solutions are not in harmony with their theoretical context, and lie outside the range of application of the field equations of GTR.

(10) Consequently, although we need not deny the usefulness of such solutions, we can deny that absolute$_1$ space–time is an irreducible element of GTR; for the theoretical context of GTR cannot be divorced from cosmological considerations.

(11) Indeed, we cannot argue that *our* space–time—which is, after all, the principal concern of GTR—is absolute$_1$.

We should note that step (9) of the relationist argument introduces a new consideration into our discussion of the theoretical context and the range of application of a law. We maintained above that if a physical situation was in conflict with the theoretical context, then the description of that situation lies outside the range of application of the law. When we are dealing with negative kinetic energy or negative absolute temperature, the conflict is decisive, and we may rightly maintain that even the *use* of the law to describe such situations is unacceptable. But when we are dealing with phenomena which we already know exist, such as rotating stars, then the *use* of the law may be acceptable, especially if it gives us good empirical results. But if we also know that this use of the law conflicts in some fundamental way with our theoretical understanding of the physical world, then we need not grant full physical significance to the results and the implications of the results obtained.

Although I believe my account of the relationist response to the absolutist$_1$ to offer the best chance of the relationist succeeding in the debate, I have two main worries about this response. These worries arise from the fact that the relationist seems to rely on the claim that we can identify a fundamental conflict between absolutist$_1$ models and the theoretical context of GTR. This claim is supported by the belief that we have strong empirical evidence for the cosmological aspects of that context. In addition the relationist argument relies on the idea of a theoretical context as a relatively stable organism, which enables him to identify both the range of application of the field equations and the physical situations which do not fall within that range.

There seems to be an underlying assumption in the relationist argument that we can specify the contents of GTR and its theoretical context. And so long as we have a neat and tidy text-book picture of GTR in mind, and so long as we do not question the empirical evidence for cosmological and other models too seriously, we may grant them this assumption. But I believe that if we look at GTR and its context in earnest we will discover tremendous debates and numerous disputes about the contents and implications of the theory. This should shake the relationist's confidence in his assertion that GTR vindicates MP. But it should also shake the absolutist's conviction that GTR requires absolute$_1$ space–time. For he too relies on a neat and tidy view of GTR to provide him with his ammunition for the fight against relationism. In the course of the next two chapters I hope to show that GTR is not all that a cursory reading of a graduate text-book might make it seem to be.

A portrait of Stephen Hawking by Yolanda Sonnabend (1985). Courtesy National Portrait Gallery, London.

Chapter 4

Classical and Quantum Relativity

Introduction

Recent work by Geroch, Hawking, Penrose and others has done a great deal to elucidate the causal structure of GTR space–times. In the course of this chapter we shall review this work and examine its implications for our view of the contents and overall structure of GTR. We shall find that we are faced with a choice: either we may decide to argue that GTR is essentially a 'classical' theory and that 'quantum' physics should have no part in the theory, or we may argue that GTR is now a blend of classical and quantum approaches to gravitation, and that purely classical methods are valuable inasmuch as they tell us how to deal with 'normal' physical situations. The choice is essentially one between maintaining that GTR is deterministic and contending that it is indeterministic. How we choose will have inevitable repercussions upon our view of the scope and contents of GTR; therefore our decision will be a major factor in determining the nature of GTR's theoretical context.

4.1 Space–time structure and singularities

In this section I shall present a critical review of recent topological studies in GTR insofar as they relate to the question of causality. Excellent technical accounts of this work are provided by Hawking and Ellis (1973), Misner *et al* (1973: ch. 34) and Geroch and Horowitz (1979). Useful background surveys are provided by Sklar (1974: chs. II, IV), Hawking and Israel (1979) and Penrose (1979).

In §3.2 we observed that if we are to have exact solutions of the field

equations then we must impose two important restrictions upon the energy-momentum tensor $\boldsymbol{T}$, i.e. upon the matter field. These are (i) a local causality condition such that the curve along which a signal travels between two points is either null or time-like and (ii) an energy condition such that the kinetic energy in a given locality is always non-negative. The immediate implication of the local causality condition is that we can only say that two points, p, q, of the space–time manifold $\boldsymbol{M}$ are causally related if we can connect them with a non-space-like curve. We shall now fill in some of the detail involved in this condition, and this will lead us on into more complex topological issues. Condition (ii) involves quite different considerations to which we shall turn in the next chapter.†

We require first the concept of *time-orientability*. We shall call a space–time M,g time-orientable when it satisfies the following definition:

D1 A space–time is time-orientable if it is possible to define continuously a division of non-space-like vectors into two classes, which we shall arbitrarily label future-directed and past-directed (see Hawking and Ellis (1973: 181)).

The consequence of this definition is that we cannot transport a null or time-like vector around a closed curve in M,g and reverse its time direction when the space–time is time-orientable, i.e. the time direction at every point is fixed.‡ We will of course have some good physical reasons to support our choice of time direction assigned to any particular point p. Thermodynamic considerations are often taken to be fundamental to our choice here since it is in the macroscopic realm of thermodynamics that time asymmetry is perhaps most striking (see Swinburne's discussion of the second law of thermodynamics (Swinburne 1968: 273–5) for an example of this approach). However, Penrose has suggested that we may be able to trace the reasons for the observed macroscopic time asymmetry back to the local conditions which obtained just after the initial singularity. He believes that as we approach the 'big bang' the Weyl curvature might tend to zero, and that the gradual increase in this component of the curvature of space–time as we move away from the initial singularity provides us with an underlying time asymmetry to which all others could then be related. Thus investigations into the laws which govern the physics of singularities might help us to deliver a

†Condition (ii) is, however, an essential ingredient of the singularity theorems which will be considered in this chapter.

‡In general, any space–time which is not time-orientable has a covering space which is (see Hawking and Ellis (1973: 181)).

clear-cut resolution to the problem of time asymmetry. If Penrose's intuitions prove to be well founded then

> the problem of time's arrow can be taken out of the realm of statistical physics and returned to that of determining what are the precise (local?) physical laws. (Penrose 1979: 633)

But until then we cannot be too sure of our choice for the direction of time through any given point of space–time. For, with a single notorious exception, all local physical processes and the local laws which govern them are symmetrical in time. The exception is the behaviour of the K^0-meson, although some writers are worried about possible temporal asymmetries in quantum mechanics (see Davies (1974), Penrose (1979) and Healey (1981)). Consequently, the choice we make for the direction of the arrow of time through a given point of space–time seems to be a matter of convention. Hence in GTR what constitutes the future and what constitutes the past of a given point turns on the arbitrary decision referred to in definition 1 (D1) above. Now if we did transport a non-space-like vector around a closed curve only to discover that its time orientation had reversed we would find ourselves in difficulties if we try to draw a global distinction between the forward and backward light cones of events in the space–time concerned. For what was to the future of a point before the transportation would be in the past afterwards. In short, in such a space–time the concepts of forward and backward in time are not well defined. Therefore we shall assume that the space–times we deal with in the following discussion all have fixed time orientations; without this assumption we would not be able to say of a given point that 'these events' are in the past and 'those events' are in its future unambiguously.

We can now move on to the concept of *precedence*. First, we divide the set of time-like vectors at p into two classes: past and future, thus choosing a direction for the arrow of time. Because the space–time has a fixed time-orientation, we can make this division at all p in ***M*** and continuously so over the manifold ***M***. So we define 'p precedes q' as follows:

D2 There is for any p,q in ***M*** a future-directed non-space-like curve which begins at p and ends at q (see Geroch and Horowitz (1979: 232)).

It is a matter of discretion whether we include future-directed null curves in D2: Geroch and Horowitz point out that what little we gain physically from the addition is outweighed by the complexities introduced into the mathematics. But for our non-mathematical discussion we shall include null curves in D2 as above.

We shall now tie causality in with the concept of precedence; we can do this by introducing a few more definitions:

D3 A causal curve is any smooth curve which is nowhere space-like.

D4 The event p causally precedes the event q if there is at least one future-directed causal curve from p to q.

D5 The causal past of p is the set of all events which causally precede p.

D6 The causal future of p is the set of all events which causally follow p, i.e. which are causally preceded by p.

These definitions are based on Misner *et al* (1973: 922–3).

It will be obvious that D3 and D4 provide us with a restatement of D2 in causal terms—simply a shift from precedence to *causal precedence.* We should note that the less obvious shift from talk of points in D2 to talk of events in D3–D6 is innocuous. As Misner, Thorne and Wheeler point out, in GTR we normally refer to the points of space–time as events (1973: 6). The use of the word 'cause' invokes a whole collection of prejudices which we would do better to forget for the moment. So when we say that p causes q we shall mean that p causally precedes q in the sense of D4. This is obviously wider in scope than the every-day use we make of the idea of causation. For example, we normally pick out certain events as being *the* cause or causes of a later event, and leave out others even though they causally precede the later event. We have certainly imported some bias into our definitions of causal precedence, namely the view that causation is a one-way process from past to future in the sense of D4. This leaves open the possibility of someone being able to influence the past by travelling along a future-directed but closed time-like curve, as we shall see shortly. But it rules out the possibility of influencing the past by travelling along a past-directed curve, i.e. of 'backwards causation'. Again, we shall try to see whether this is justified later. We have now unpacked the basic ideas involved in the first 'local causality' condition we impose on the field equations; but, as we shall see, a good deal more can be said about the causal structure of space–time without making many more assumptions.

A number of causal anomalies due to the physics of curved space–time are fairly well documented: they are (i) closed time-like curves, (ii) almost closed time-like curves, and (iii) closed null curves. Briefly, a closed time-like curve is a time-like curve from a point p through distinct points r, s, t, . . . but eventually reaching the source point p. An almost closed time-like curve is a time-like curve from p through r, s, t, . . . but eventually reaching a point w such that w is arbitrarily close to p even though the curve itself does not reach quite

as far as p, and therefore through an arbitrarily small neighbourhood the same time-like curve will pass more than once. A closed null curve is a null curve from p through r, s, t, . . . but which returns to p. In each of these cases there seems to be the possibility of someone at p influencing his own past by sending a signal or perhaps himself along the curve from p. What influence a person could have upon his past by 'going forward into the past' is a moot point. A number of so-called paradoxes have been constructed to counter suggestions that travel along such curves is plausible or even possible in principle. Many of these involve the liberal use of psychological concepts like 'free will', 'action', 'belief' and so on. I agree with Horwich (1975) that we should not be surprised if these concepts, which are designed for and work well in ordinary situations, break down in 'the bizarre context of time travel' (Horwich 1975: 438). Some paradoxes seem to pose a greater threat to closed time-like curves etc: these do not rely on the above psychological concepts and generally involve the presence of two incompatible events on the same closed curve. For example, there is the paradox of 'auto-infanticide': someone returns to his past and murders his infant self, but since an unsuccessful attempt is a necessary condition for a successful attempt on the infant's life, the explicit contradiction here leads us to the conclusion that this mode of time travel is impossible. There are other examples along the same lines but which merely involve the transmission of signals or the sending of inanimate objects (see Earman (1972)). However, Horwich points out that the paradoxical appearance of such examples derives from the following source:

> No conceptual difficulties are involved in the idea of a causal chain in which someone goes on a journey and kills someone at his destination. Problems arise only when we consider these causal chains to be located along closed timelike curves; . . . This view embodies the idea that . . . if some causal chain can be located along some timelike line, then it should be locatable along any sort of timelike line—even a closed one. (1975: 442–3)

But, as Horwich goes on to say, it is wrong to maintain that any causal chain can be located along any sort of time-like curve. For

> . . . any causal chain must be consistent with those chains which are located along intersecting timelike lines . . . [i.e. it] must satisfy consistency conditions imposed by its surroundings. (1975: 442–3)

We can be non-committal about the meaning of 'causal chain' here: all we need to say is that it is a continuous sequence of events along a continuous space–time curve. So Horwich is claiming that the events

on a time-like curve should be consistent with the events in the locaility through which the curve passes—and this seems reasonable. Now examples like the 'auto-infanticide' case produce outrageous inconsistencies when we try to fit the sequence of events they require onto a closed time-like curve—not only in the sense that the stories we tell about such cases involve internal contradictions but also in the sense that we cannot swallow these stories without compromising their 'external' consistency with the context in which they are set. Horwich seems to believe that the basic test for the admissibility of a particular closed time-like sequence of events is that of external consistency. But we need to show—and Horwich does not—that a sequence which passes this test could not be internally inconsistent. Fortunately, this is not too difficult. Consider a closed time-like curve through points A, B, C, . . . A and a time-like curve P′, Q′, . . . which intersects the closed curve at B (see figure 4.1). Now whatever we say about the closed curve is going to be constrained by the curve through B: the event at B will have to remain precisely the same, since it is an event on the P′Q′ curve. But B depends on what happens at A, and this depends on what happens at all the points back to C. So once we have fixed B in the closed curve we fix all the points on the curve. And we can accept that the curve is consistent with its environment which we have represented by the open curve. What happens when we try to introduce an internal inconsistency? Let us use a variation of an example by Earman (1972) as follows. There is a switch at B which is in the 'on' position; a signal is sent from C forward to A if and only if the switch at B is 'on'—this signal is designed to move a lever at A which throws the switch at B into the 'off' position. But if the switch is 'off' the signal from A cannot be sent which means the level at A is not moved which means the switch is 'on'. Hence the example involves an internal contradiction, or rather a series of contradictions: the signal is sent and not sent; the lever is both moved and unmoved; and the switch is 'on' and 'off'. But if the time-like curve A, B, C, . . . A is to be externally consistent then as we have observed B is fixed. So we can only tell stories in which the switch is either 'on' or 'off' but not both. And which position it is in fixes the other events in the sequence. Consequently, if the curve P′, Q′, . . . demands that the switch at B is 'on' we can then have only sequences of events along the closed curve A, B, C, . . . A which are consistent with this demand. And the above sequence is certainly not consistent with it. Given that we can intersect the closed curve with another curve at any point in the sequence A, B, C, . . . A, what happens at the point of intersection will be fixed. Hence total consistency with all intersecting time-like curves implies that there is no possibility of any ambiguity about the status of any event on the closed curve. This rules out all curves with internal

inconsistencies. Therefore we can accept Horwich's assertion that so long as a closed time-like curve is externally consistent we cannot rule it out as a physical possibility.

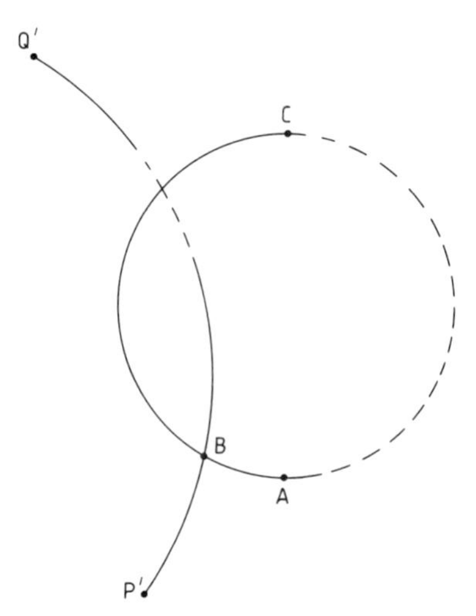

Figure 4.1 Intersection with a closed time-like curve.

Are closed time-like curves etc, more than just a mere possibility? In other words, are we justified in giving them serious consideration? Geroch and Horowitz think that we ought to do so because

> physical theories frequently suggest new and unexpected phenomena, which are later found to be realised physically. (1979: 238)

A famous example of this is the singularity of collapse first put forward in a relativistic context by Chandrasekhar in 1932 (see Mehra (1973: 36–7)).† As Chandrasekhar says (in Mehra) the questions he then raised concerning the indefinite compression of matter 'failed to attract any attention'. More to the point, the only attention attracted was adverse: Eddington thought that the idea was an absurdity (see Mehra (1973: 38)), and the rest of the scientific community followed his lead. But Eddington appears to have been hasty given the recent work of Geroch, Hawking and Penrose. The culmination of this work, the Hawking–Penrose theorems, imply that

†The first time the possibility of such a phenomenon seems to have been considered was by Laplace in 1798—this paper's English translation appears as Appendix A in Hawking and Ellis (1973).

> . . . a singularity is inevitable in the gravitational collapse of a star once it has passed a certain critical stage. This point of no return occurs when the gravitational field becomes so strong that it drags back any further light emitted by the star. (Hawking and Israel 1979: 15)

This case study in scientific inertia could well be regarded as a salutary warning to those who follow Eddington's example on the question of closed time-like curves (see for example North (1965: 38) and Reichenbach (1971: 139)). However, the number of 'particles' 'discovered' during the early days of particle physics should stifle any undue enthusiasm for believing without reservation in every new phenomenon which comes along, for a number of those particles cannot be taken too seriously any more.† So we obviously need to keep an open mind on phenomena like the problematic curves we have been dealing with; but we should not let things get out of proportion. There is little or no empirical evidence to suggest that closed time-like curves are anything more than mathematical curiosities. But they are not theoretically otiose, and we certainly do not have *a priori* grounds for ruling them out. Indeed, there seems to be a good deal in common in the cases of closed time-like curves and singularities of collapse, at least when we compare their early theoretical history. If anything, closed time-like curves have rather more theoretical respectability now than black holes had until quite recently. Nevertheless, I shall follow the present day convention and require that GTR space–times should be free from the curves under discussion. We can do this by imposing the condition of *stable causality* upon space–time.

We can define stable causality as follows:

D7 A spacetime is causally stable if (i) there are no closed time-like curves and (ii) we can expand the light cones slightly at every point without introducing closed time-like curves.

This definition is based on Geroch and Horowitz (1979: 241) and Hawking and Ellis (1973: 189–98) who show that all the causally anomalous curves above violate the stable causality condition—this is a technical matter which need not concern us other than by noticing that when we open out the light cones more curves become time-like and consequently the chances of finding a closed time-like curve are increased. Stable causality also implies the assumption we made earlier that we can distinguish between the past and the future of each point of space–time, i.e. that the space–time is time-orientable.

Apart from ruling out space–times with the class of causal anomalies described above, there are two important features of stable

†See Pais (1986: ch. 20), who describes how the apparently distinct χ, κ and τ particles were shown to be a charged K^+ particle.

causality which we should note. We shall find that the first is of crucial significance when we come to discuss the structure of GTR in the next chapter; the second will be of direct concern now.

(1) Hawking (1966) has shown that we can determine the differentiable and conformal structures of the manifold from the causal structure implied by the condition of stable causality. Hence there is reason to believe that this causal structure is a fundamental feature of the manifold, certainly more basic than the differentiable and conformal structures which are themselves usually taken to be more basic than the affine and metric structures (see Sklar (1974: 46–54) and Friedman (1983: 139*f*)).

(2) In a causally stable space–time there is a time function f on the manifold such that f increases along every future-directed non-space-like curve. Whereas time orientability gives us a local time sense allowing us to distinguish between past and future locally, the time function provides us with a global time sense for a space–time.

The most striking feature of the time function f is that it enables us to construct '*time-slices*'; Hawking and Ellis tell us

> The spacelike surfaces (f = constant) may be thought of as surfaces of simultaneity in spacetime, though of course they are not unique. (1973: 201)

The appropriate physical notion for the surfaces defined by f = constant is that of 'all space at an instant of time' hence the appellation 'time-slice' (see Geroch and Horowitz (1979: 243)).

All causally stable space–times have time-slices passing through every point of the manifold. However, in general there will be more than one slice through each point. This is because there will normally be more than one set of time-slices defined by 'f = constant' for a given space–time. For example, in the causally stable Minkowski space–time we can define the time-slices

$$\text{the plane } t = 0$$
$$\text{the hyperboloid } t = (x^2 + y^2 + z^2 + 1)^{1/2}$$

where the Minkowskian coordinate t is a time function—hence the lack of uniqueness referred to by Hawking and Ellis. Besides Minkowski space–time there are a number of important causally stable space–times; e.g. Schwarzschild, Friedmann and Reissner–Nordstrom. Amongst the space–times which are not causally stable are those of Gödel and Taub–NUT. The Gödel space–time is perhaps the most notorious of those space–times which admit closed causal or time-like curves (see my remarks in §3.2(c)).

Whitrow (1980) discusses the relationship between GTR and what

he calls 'cosmic time'. The idea of time-slices in causally stable space–times seems to allow for the possibility of a universal time-scale. Hence, he argues that GTR is compatible with a universal 'absolute' time frame. Whitrow maintains that the observed homogeneity and isotropy of the universe are evidence for the existence of cosmic time. Certainly, many cosmologists sometimes regard the Friedmann solutions as providing such a time-scale. However, as pointed out above, we can normally produce more than one sequence of time-slices in causally stable space–times. So, in general, no unique cosmic time can be defined. The use of any such time-scale must be instrumental and cannot imply the ontological status sought for it by Whitrow.

A concept closely related to that of the 'time-slice' is the *achronal slice*.

D8 A slice is achronal if no point on that slice causally precedes any other point on that slice (see Geroch and Horowitz (1979: 246)).

Again, we may think of an achronal slice as a surface of simultaneity in the space–time with which we are concerned. The slice need not consist of *all* space at an instant, but when it does we may refer to the achronal slice as a *hypersurface*. Just as causally stable space–times admit time functions, they also admit achronal slices. As Geroch and Horowitz point out

> . . . since the time function is increasing along timelike curves, any slice of constant time function must be achronal. (1979: 245–6)

Indeed, the plane and the hyperboloid of Minkowski space–time are examples of achronal slices. When an entire plane or hyperboloid is considered we will of course be dealing with a space-like hypersurface. The major difference between time-slices and achronal slices lies in the introduction of the concept of causal precedence into the definition of achronality. This allows us to specify that for a given achronal slice there will be no causal curves between any two points of the slice. Causal curves pass through the slice from points which precede it to points which it precedes. Obviously, for a given causally stable space–time, the set of achronal slices as defined by a particular constant time function provides us with layers of space-like surfaces (see figure 4.2).

Causal curves will pass from slice to slice. We should recall that we can define our time-slices in various ways as we saw in the case of Minkowski space–time. Hence there is no immediate reason why we should grant any one set of achronal slices which are hypersurfaces a particular physical significance, e.g. by saying that one set rather than

another represents *the* surfaces of simultaneity for a space–time.† The significance of the achronal slice resides in the fact that since causal information passes through it along the causal curves we can specify the physical situation at a given instant by saying exactly what is happening on the slice, and this will be related via the causal curves to past and future slices.

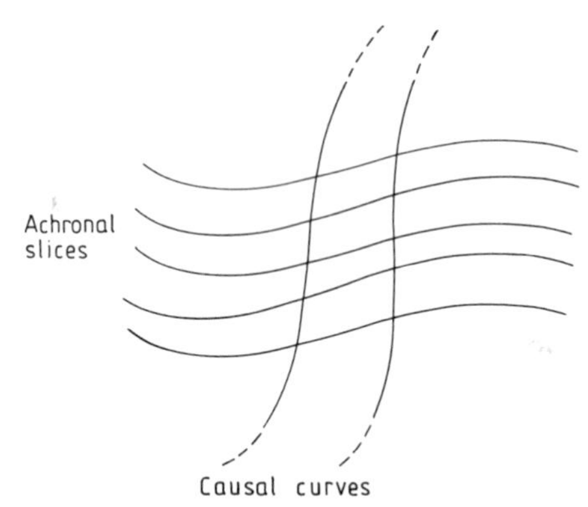

Figure 4.2 Causal curves and achronal slices.

The relationship between an achronal slice and the points to its past and future can be made precise via the introduction of the concept of *domain of dependence*. Consider a point p in a causally stable space–time and the set of all non-space-like curves passing through p and through an achronal space-like region S to the past of p. So long as every past-directed curve from p cannot be assigned an end-point, we can define the future domain of dependence of S, $D^+(S)$, as follows:

D9 $D^+(S)$ is the set of all points p such that every past-directed non-space-like curve from p meets S.

So, only curves from S can reach and therefore have a causal influence on p. Hence p depends entirely on what is happening on S, i.e. p is causally determined by S. Whilst a point q outside $D^+(S)$ may be causally preceded by one or more points of S, q cannot be determined by S since causal curves from points on the same space-like time-slice as, but lying outside of, S can reach q. These details are made clear in figure 4.3.

Similarly, we can define the past domain of dependence of S, $D^-(S)$, as follows:

†Later (p.148*f*) we shall consider the claims of one such set for special status.

D10 $D^-(S)$ is the set of all points p′ such that every future-directed non-space-like curve from p′ meets S.

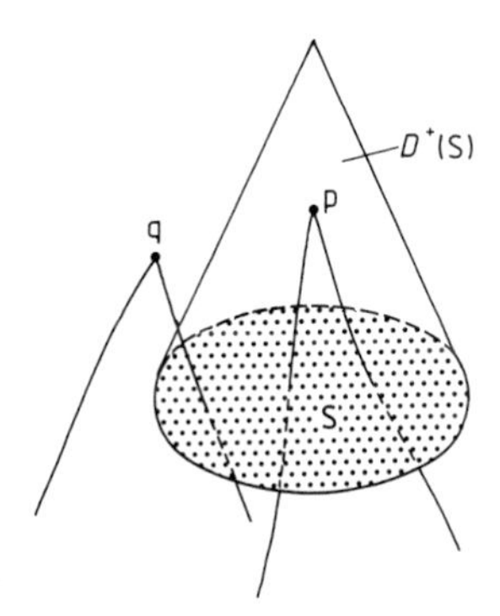

Figure 4.3 Domain of dependence.

Since all curves from p′ reach S what is happening on S uniquely determines the physical behaviour of p′. I do not mean in either of these cases of determination that what is happening on S *causes* p and p′. Whether this is so needs to be considered. The notion of causal determination contains no more at this point than the ideas spelt out in our definitions so far, that is an event at p or p′ is causally determined by the events on S only if p or p′ are in, respectively, $D^+(S)$ or $D^-(S)$.† Consequently, causal determination is indifferent to time direction, contrasting with causation which, we have agreed, goes from past to future.

To find out precisely what is going on in both $D^+(S)$ and $D^-(S)$, i.e. in the domain of dependence of S or $D(S)$, we need to utilise the field equations of GTR. This is a far from trivial exercise. Essentially, given all the relevant information concerning the physical situation on S—*the initial data set* on S or IDS(S)—we can find a unique space–time which satisfies the field equations, has an achronal slice S, provides exactly the same data on S as specified in IDS(S), and is such that the space–time lies entirely within $D(S)$. See Geroch and Horowitz (1979: 250) for a more detailed description of the operation. The consequence of this rather involved exercise is the determination of all events in $D(S)$ by IDS(S). However, the possibility of this determination rests upon an assumption made earlier, namely that past-directed curves from p and similarly future-directed curves from p′ can always reach S, i.e. we cannot assign an end-point tc any such curve before it reaches S. We should note that since we can always smooth out the

†Determination of closed causal curves to future and past of S should be possible, so long as they lie entirely within $D^+(S)$ or $D^-(S)$ respectively.

'join' when a curve from the past meets a curve going to the future at a point p, we may consider these two curves as a single curve through p. Consequently, when a curve has an end-point at p there will be no curve from p in the opposite time direction. How could this arise in GTR space–times? And do we have any reason to legislate against end-points in the sense described? Otherwise we could not rely on the IDS alone when we try to determine what is happening throughout *D*(S), or rather the region of space–time we would expect IDS(S) to determine. Such regions we shall denote by *D*(S)*, and we shall do so whenever we have reason to suspect that 'total' determination might be disrupted.

The simplest way to demonstrate the difficulties created by curves with end-points is to utilise Minkowski space–time with one or more points removed at the origin, i.e. at 0,0,0,0. If we try to link the points p (at −1,0,0,0) and q (at 1,0,0,0) with a non-space-like geodesic of maximal length pq—the straightest curve from p to q—we will discover that we cannot do so. In effect, the removal of the point(s) cuts a hole (topological not black of course) in the space–time manifold and this blocks the path of the geodesic pq. Of course, there will be non-space-like curves from p to q which avoid the hole, but any curve which meets such a hole will have an end-point in the sense described above. In our example, shown in figure 4.4 below, a past-directed curve from q and a future-directed curve from p both meet the hole, which prevents the two curves from meeting each other and therefore rules out the possibility of us smoothing out any join between them. Quite simply, a line (or curve) with a hole in the middle is really two distinct lines (or curves). And this is what happens to the curve pq when we cut a hole as depicted in figure 4.4. If p is in the causal past of the hole h, i.e. J^-(h), and q is in the causal future of h, i.e. J^+(h) (or vice versa) then there can be no continuous curve pq *via* the hole, no matter where we position p and q in J^+(h) and J^-(h) (see figure 4.5).

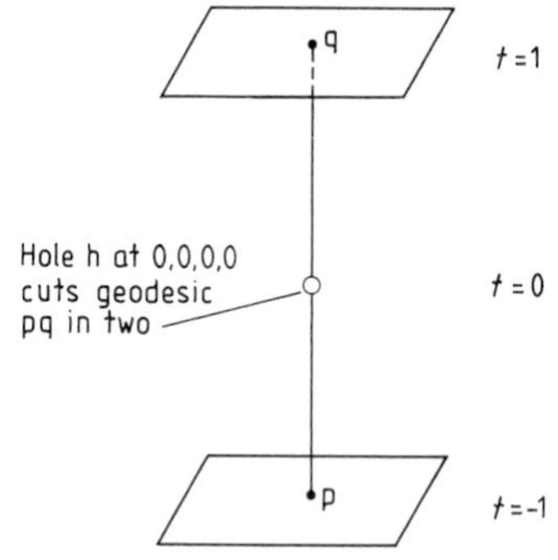

Figure 4.4 A topological hole.

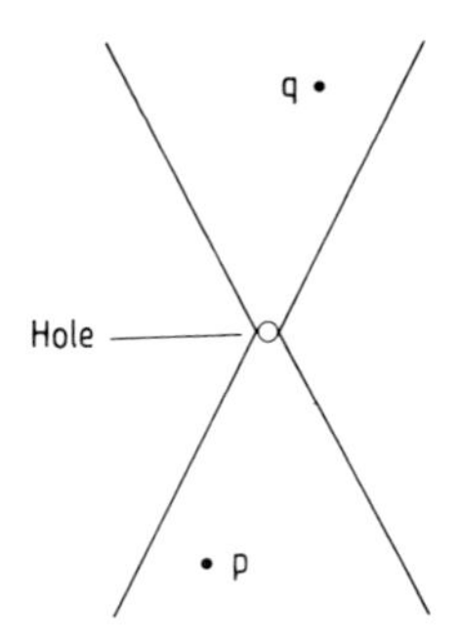

Figure 4.5 Effect of a topological hole.

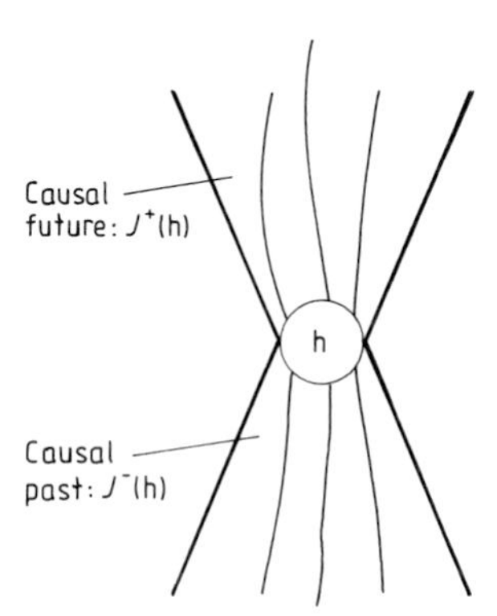

Figure 4.6 Causal future and causal past of a topological hole.

We should note that, strictly speaking, we should refer to the causal future and past of the points around the edge of h rather than of h itself, since by definition a hole is an absence of points. These points around the edge of h represent the end-points of curves. Some of these non-space-like curves will be only to the future of h, and the rest only to the past (see figure 4.6).

The effect of the hole is to cast a 'shadow' which interferes with the causal determination of events which lie within its bounds. This is because non-space-like curves may start at the hole and affect events in the causal future of the hole J^+(h), with the result that an achronal surface S to the past of h cannot determine events in J^+(h) in the shadow region. Certainly, events on S may influence events in J^+(h), since curves from S will find their way into this region. But the information given in the IDS(S) will give us no idea of what curves will emerge to the future of h from h; indeed, there will be no indication in IDS(S) that curves from S will meet h—topological holes always take us by surprise. Consequently, IDS(S) does not determine the events in J^+(h) even when they lie in D^+(S)*. This situation is depicted in figure 4.7

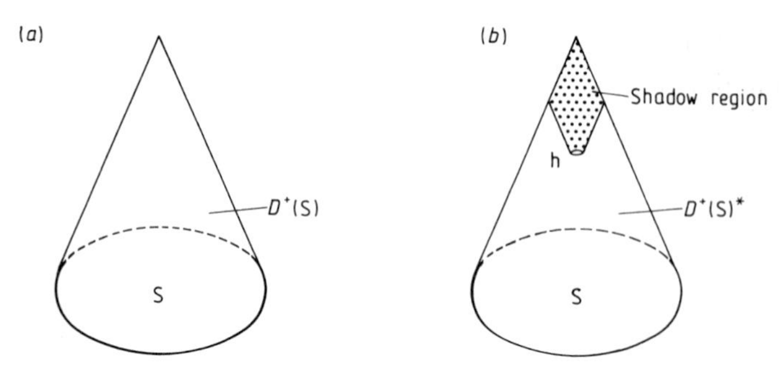

Figure 4.7 Future domain of dependence and shadow region.

Similarly, curves may end at h so that an achronal surface to the future of h cannot determine events in $J^-(\mathrm{h})$. Once again, there is no way of telling from IDS(S) what the effects of h might be and indeed whether h is present in $D^+(\mathrm{S})^*$ (see figure 4.8).

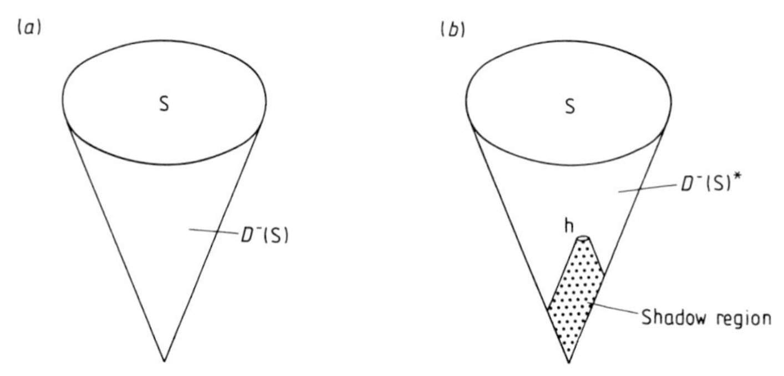

Figure 4.8 Past domain of dependence and shadow region.

The consequence of allowing topological holes in the space–time manifold is that we can never be certain that the events we believe to be determined by a given IDS(S) are actually so determined. Holes interfere with causal determination. Consequently, if we are to rely on any IDS, we must legislate against holes in the manifold. Have we a reason to do so? Geroch suggests that since space–times with topological holes have no known practical application, we could modify GTR to exclude them without any real loss (see Geroch (1977: 87)). But even if we agree that holes in space–time are no more than topological fantasies, the violations of causal determination they produce can also occur when space–times contain certain kinds of singularities. And although cutting a hole in the manifold might seem a rather artifical device, singularities now have a good physical pedigree.

What follows is a very brief account of some of the problems raised by singularities. This area of GTR is very complex indeed, so it will soon become clear why I chose the relatively simple idea of topological holes as an introduction to the violation of causal determination.

It is not easy to give a neat definition for singularities in GTR. Usually, we think of singularities as points of space–time where some physical quantity, e.g. charge, is infinite. But in GTR if matter is 'squeezed' to become infinitely dense, then the curvature at the point of infinite density will be infinite. But, as Geroch and Horowitz tell us, the field equations of GTR are not able to take account of such space–time singularities:

> General relativity, as it is usually formulated, requires a manifold with a smooth Lorentz metric. This formulation leaves no room for points of the manifold at which the metric is singular. Indeed, it is hard to see how one could modify the theory to admit such "singular points", for it is only through the metric that one acquires the ability to identify the individual points of the manifold as events. One cannot isolate, as additional physical events, points at which the metric is badly behaved. In short, it seems to be a necessary part of general relativity that all "singular points" have been excised from the spacetime manifold. (1979: 256–7)

We have already mentioned that Hawking and others have shown that under certain conditions singularities are inevitable. How does this square with the recommendation above that we should not consider space–times with singularities? Our first reaction might be that singular points just never happen—that some physical process might occur at very high densities which prevents the actual collapse (or whatever) to infinity. But, as Sciama points out, this does not help us:

> . . . in general relativity self-gravitation is so strong that even in our irregular universe some at least of the material must have been squeezed to infinite density. Quantum mechanical considerations might possibly enable us to avoid a literal singularity but apparently they cannot prevent the density getting very high indeed, say 10^{59} g cm^{-3} . . . For practical purposes such a density might well be regarded as a singularity. (1971: 127)

Even so, the idea that we might be able to avoid a singularity has little as yet to support it.

To prevent the embarrassing breakdowns in GTR which singularities seem destined to produce, most physicists accept Penrose's

remedy: that a 'cosmic censor' is at work.† Hawking and Israel describe how the idea of cosmic censorship helps GTR to keep its head above water:

> A cosmological singularity in the past is bad enough because it implies that there was a time before which nothing was defined and no questions could be asked, but a singularity in the future, such as is predicted to occur in gravitational collapse, is even more disturbing because it means that time itself comes to an end as would any helpless observer who was foolish enough to follow the collapse of the star. Near the predicted singularity the gravitational field would presumably get very large and as yet unknown quantum gravitational effects would be expected. Thus the singularity theorems show that classical general relativity predicts its own breakdown. However it seems that we may be shielded from the full effects of a breakdown because it may always occur in a region of space-time, a black hole, which is not visible to an external observer, at least classically. This conjecture was given the name "the cosmic censorship hypothesis" by Penrose. Without it we could be in for a surprise every time a star in the galaxy collapsed. (1979: 16)

So, in the case of a singularity of collapse which is passed by the cosmic censor (whom no one supposes actually exists!), no external observer can detect precisely what is happening at and close to the singularity. For the singularity is surrounded by an absolute event horizon through which light and other particles can fall but from which none can emerge, at least by classical means. Our knowledge *qua* GTR of the internal region of a black hole is limited to its mass, angular momentum and charge—in fact nothing which might upset our comfortable GTR view of what is going on in space–time. To say what is happening inside the event horizon we would probably need a quantum theory of gravity—or some novel theory—and so far no convincing account has been advanced (see Hawking (1980)). Consequently, the event horizon saves GTR from trying, and failing, to provide an account of what is happening at and close to singularities. But how reasonable is the cosmic censorship hypothesis? Does it mean that all space–time singularities must be black holes? And are black holes as innocuous as the cosmic censor would have us believe?

If we are to answer these questions, we will have to say a little more about singularities. We have already observed that there is no neat definition of singularities in GTR and so far we have spoken rather

†Penrose (1979: 617–29) gives the details of his hypothesis together with a general defence. See also Wald (1983: 302–5) who presents and discusses various formulations of cosmic censorship.

loosely of points of infinite density. But, as Geroch and Horowitz point out, to characterise singularities as points at which components of the metric become infinite or display other signs of bad behaviour is not appropriate. There are two reasons for this.

(1) We can create such singular points by choosing our co-ordinates unwisely; and it is difficult to separate 'genuine geometrical information about the spacetime from "gauge" information contained in the chart-choice' (Geroch and Horowitz 1979: 256).

(2) It is not at all clear what 'points' are being referred to when we speak of such singular points (Geroch and Horowitz 1979: 256).

In short, if we are to have singularities, then we want them to be genuine and identifiable physical phenomena rather than artifical constructions. It turns out that the only way we have of making the notion of a physical singularity precise is to define a singularity in terms of geodesic incompleteness.

D11 A space–time is geodesically incomplete if it contains one or more maximally extended non-space-like geodesics whose affine parameter does not assume values in the full range from $-\infty$ to $+\infty$.†

This definition is based on Geroch and Horowitz (1979: 257). Hawking and Geroch suggested that this definition provides the basis for a definition of singularities. In the words of Hawking and Ellis:

> Timelike geodesic incompleteness has an immediate physical significance in that it presents the possibility that there could be freely moving observers or particles whose histories did not exist after (or before) a finite interval of proper time. This would appear to be an even more objectionable feature than infinite curvature and so it seems appropriate to regard such a space as singular. Although the affine parameter on a null geodesic does not have quite the same physical significance as proper time does on timelike geodesics, one should probably also regard a null geodesically incomplete spacetime as singular both because null geodesics are the histories of zero rest-mass particles and because there are some examples (such as the Reissner-Nordstrom solution) which one would think of as singular but which are timelike but not null geodesically complete. As nothing moves on spacelike curves, the significance of spacelike geodesic incompleteness is not so clear. We shall therefore adopt the vew that *timelike and null geodesic completeness are minimum conditions for spacetime to be considered*

†The affine parameter along a time-like geodesic is a measure of proper time.

> *singularity-free*. Therefore [**D12**] if a spacetime is timelike or null geodesically incomplete, we shall say it has a singularity. (1973: 258)†

And we shall call a space–time with a singularity simply geodesically incomplete or GI. We are now in a position to define a singularity:

D13 A singularity is a physical phenomenon which brings about a GI space–time.

Loosely speaking, it is a 'place' where geodesics simply terminate. I have added the stipulation that the singularity should have a *physical* basis bearing in mind the ease with which 'singularities' could be created by an ill-advised choice of coordinates. Although basing our definition on the concept of GI space–times escapes that particular problem, we shall now see that a quite different 'artifical construction' can produce a GI space–time.

The view that singularities are 'places' on the manifold where geodesic curves terminate takes us onto familiar ground. We have already seen that topological holes are 'places' where geodesics can go no further, where curves have their end-points. Quite naturally, this fact suggests the question of whether topological holes are singularities. We should also ask whether singularities interfere with causal determination in the same way as topological holes.

Certainly, in all respects but one topological holes seem to be singularities. For they do give rise to time-like and null GI space–times. But as we observed above, cutting a hole in the manifold does not seem to have any physical basis. Of course if we do discover that such holes have a physical pedigree, then we could rightly call them singularities. But the point made by many physicists is that until we have at least some evidence for taking topological holes (and other geometrical oddities) seriously, there is no need to complicate our ideas about the physics of GTR space–times. Singularity theorems are complex enough already. I believe that the real value of topological holes lies in the fact that they are conceptually useful devices, despite their artificiality. They allow us to focus upon certain properties of

†I am avoiding a fairly technical matter here which bears mainly on D13. There are space–times which physicists wish to call singular inasmuch as what goes on in them has the appearance of geodesic incompleteness, but which are not actually GI (see Geroch (1968)). Hawking and Ellis show how we can take account of these space–times by adopting the notion of bundle completeness which implies geodesic completeness: the essential idea involved is that a space–time is singular if curves with finite lengths have no end-points (see Hawking and Ellis (1973: 259–61)). But we shall act as if D13 catered for these problem space–times. As long as we bear in mind that we should really be widening our definitions of singularities to include bundle completeness nothing will be lost, but simplicity will be gained.

space–time, particularly those which involve causal structure, and they also play an important role in checking various statements and theorems about the nature of space–time (see Geroch and Horowitz (1979: 212–16) for details of this role). Indeed, we have already shown how holes in the manifold help us to demonstrate the ways in which violations of causal determination can occur. With certain kinds of singularity the violation of causal determination is not always so straightforward; topological holes provide a neat paradigm for what can go wrong, and given the problems involved in saying what is happening in the vicinity of certain singularities this paradigm will give us a useful point of reference.

The problem of singularities and causal determination brings us back to the issue of cosmic censorship. The cosmic censor wanted us to rule out those singularities without an event horizon. Unless we do so we face two difficulties.

(1) If we accept that the region close to a singularity is highly curved and that quantum effects would probably dominate the physics there, the field equations of GTR would almost certainly not apply in that region. We noted above that the existence of an event horizon absolves the external observer from the responsibility of trying to give an account of the physics in the internal region. If no horizon exists, the GTR seems to be in trouble.

(2) Remembering that the main function of the horizon is to prevent particles etc leaving the internal region, the existence of curves starting at and moving away from the singularity is certain to interfere with causal determination. That curves will start at the singularity follows from D12 which implies that singularities are the end-points of curves. This interference will probably be similar to that of topological holes but is likely to be complicated by quantum effects.

As we have already noted, black holes avoid the first difficulty, and their event horizons seem to help them avoid the second also. Curves which start at the singularity do not seem able to get past the horizon and therefore they will not interfere with events outside. Indeed, the fact that black holes are predictable via the singularity theorems is a distinct advantage for the process of causal determination. This is as it should be, since the motivation for the hypothesis of cosmic censorship resides in the desire that physical theories should, as much as is possible, allow us to predict and determine events in their domains. We must now face the crucial question: has the cosmic censor got it right? Are there any reasons to suppose that singularities without horizons exist or could exist in the physical world?

We have said that most physicists accept the cosmic censorship hypothesis, but there are some forceful dissenting voices. For example, Clarke maintains that singularities without horizons—

'naked' singularities—are physically respectable (see Clarke (1977: 105)). The truth of the matter seems to be that the issue is somewhat clouded: there are GTR models which have nakedly singular space–times, but these examples seem to be physically unreasonable except in one important instance. Let us look at the exception first. The initial singularity in the Friedmann solutions is naked,† but it seems absurd to rule out this singularity. So perhaps we should be a little more precise about what it is we find problematic in naked singularities, and see if we can reasonably rule out those singularities which are badly behaved. We can best do this by saying what is right about the initial singularity.

A space–time with an initial singularity is certainly geodesically incomplete, but it is singular in an identifiable way from the beginning; as Geroch and Horowitz say (1979: 271), it has 'singular initial conditions'. Although these conditions do not have the protection of an event horizon, only the first difficulty above seems to apply. We are forced to admit that the laws of GTR will not apply at those early times when the metric is highly curved. Similarly, in closed Friedmann solutions with a final singularity—the 'big goodbye'—the last few moments will cause problems for GTR. Indeed, the final singularity is just as naked as the initial singularity, which is exactly what we would expect since the former is the time-reverse of the latter in all respects except possibly entropy level and the measure of Weyl curvature (see Penrose (1979: 633)). But although space–times with such singularities are GI, with curves starting at initial singularities and terminating at final singularities, the second difficulty above does not appear to apply. Our explanation of interference with causal determination was in terms of the effects of topological holes. If we have all the relevant information about an achronal slice to form an IDS then holes will produce shadow regions which will interfere with causal determination on the basis of the IDS. We can always determine events on our time-side of a hole, so long as they are within $D(S)$. It is the other side which causes the problems, for it is always there that the shadow region prevents causal determination throughout $D(S)^*$. So initial singularities can only interfere with the causal determination of events in their past, since we are to the future of them; and final singularities can only interfere with the causal determination of events to their future, since we are in their past. Obviously, it is not clear what such events would be anyway. But all events in 'our' space–time, as specified in the Friedmann solutions, could be determined by any observer in possession of the appropriate information so long as he does not try to determine events close to the singularities.

†Unlike some solutions with an initial singularity, the Friedmann solutions (with $\Lambda = 0$) do not possess an event horizon.

We can now define the bad behaviour which makes naked singularities undesirable. A singularity which is not evident in the initial (or final) conditions of a space–time is likely to be badly behaved. Like a topological hole it will probably occur without warning and could be difficult to detect even when it lies to our past. Like topological holes, curves will enter it and leave it without us being able to say exactly what is happening in the shadow regions. Unlike topological holes the situation is likely to be complicated and made even more indeterminate by quantum effects.

It is therefore a mistake to understand cosmic censorship simply in terms of event horizons. The principal concern of the hypothesis is with the determination of events in space–time. Of course, not being able to specify what is happening close to singularities is an embarrassment for GTR, but if we can confine the problem to initial and final singularities and to the internal regions of black holes then our difficulties are minimised.† We then have a clearly defined area within which GTR can be used effectively. For this reason we can regard such singularities as well behaved; we shall therefore reserve the prefix 'naked' for any singularity which mimics the bad behaviour of topological holes. The existence of naked singularities—just as that of holes in the manifold—would cast doubt on all calculations we make concerning any portion of space–time, no matter how distant that portion might be from better behaved singularities. Consequently, without naked singularities (and topological holes) we can safely use the information specified by an IDS to determine events in $D(\mathrm{S})$; in this case $D(\mathrm{S})^*$ *is* $D(\mathrm{S})$. These remarks lead us to a definition for naked singularities in terms of the determination of events:

D14 Given a non-singular slice S and its IDS in a hole-free space–time ***M***, if a singularity develops in $D(\mathrm{S})^*$ without an event horizon then that singularity is naked.

So $D(\mathrm{S})$ does not include any of the causal future of the singularity when it is to the future of S nor any of the causal past of the singularity when it is to the past of S—hence the use of $D(\mathrm{S})^*$ instead of $D(\mathrm{S})$ in D14. This definition is taken not to apply to initial and final singularities, since they have no identifiable causal pasts and futures respectively.

So we return to the question: are naked singularities physically reasonable? The answer seems to be a rather hesitant no. Commenting on this issue, Hawking and Israel are reluctant to commit themselves:

†However, the quantum effects associated with these phenomena still raise questions for the cosmic censor; we shall deal with these towards the end of the section.

> Despite considerable effort, the cosmic censorship hypothesis remains undecided within the framework of classical general relativity . . .. However, the hypothesis is supported by perturbation and computer calculations and by the failure of a number of attempts to obtain contradictions from it. It is therefore generally believed and forms the basis of all work on black holes. (1979: 16)

Certainly, those space–times which are said to contain naked singularities involve distinctly strange behaviour by their matter contents. We may conclude, therefore, that although naked singularities are not yet physically otiose they are some way from the physical respectability which Clarke claims for them, despite some recent success in generating such singularities from non-singular initial conditions (see footnote p.156). Hence we shall follow the crowd once again and accept the cosmic censorship hypothesis.

Before we look at some of the consequences of cosmic censorship for GTR space–times we should first mention that black holes may not be as harmless as they might seem. The event horizon is designed to prevent particles etc from leaving the internal region of the black hole. This might seem to suggest that as more and more matter falls into the hole it gets bigger and bigger until perhaps it swallows the entire matter content of the space–time. Can this happen? Hawking has shown that particle creation can be expected to take place in the following sense:

> . . . applying quantum mechanics to matter fields in the background geometry of a black hole (leads) to a steady rate of particle creation and emission to infinity. The emitted particles would have a thermal spectrum with a temperature proportional to the surface gravity of the black hole, which is a measure of the strength of the gravitational field at the event horizon and which is inversely proportional to the mass. This emission would enable the black hole to remain in equilibrium with thermal radiation at the same temperature. (Hawking and Israel 1979: 18)

The emission of radiation therefore makes the black hole into a kind of 'heat engine'—see Davies (1980: 166)—which restricts the size of the hole. Hawking and Israel suggest that

> . . . one can think of the emitted radiation as having come from inside the black hole and having quantum mechanically tunnelled through the potential barrier around the hole created by the gravitational field, a barrier that could not be surmounted classically; . . . it is possible for a black hole to emit a television set or Charles Darwin, but the number of configurations (of particles) corresponding to such exotic possibilities is very small. The overwhelming probability is for the emitted particles to have an almost thermal spectrum. (1979: 19)

Obviously, this emission, dubbed the 'Hawking process', is going to pose problems for any attempts to determine events in the causal future of a black hole. It turns out that the process introduces rather more indeterminancy than even that associated with quantum mechanics:

> In classical mechanics one can make definite predictions of both the position and velocity of a particle. In ordinary quantum mechanics one can definitely predict *either* the position *or* the velocity but not both. Alternatively one can predict one combination of position and velocity. Thus, roughly speaking, one's ability to make definite predictions is cut by half. However, in the case of particles emitted by a black hole, one can definitely predict neither the position nor the velocity. All one can predict is the *probabilities* that the particles will be emitted in certain modes.[†] This loss of predictability seems to be associated with the breakdown that one would expect at a spacetime singularity. One could interpret it by saying that new random information is entering the region of the universe that we observe from the interior of the black hole. (Hawking and Israel 1979: 19)

We should not be misled by the idea of particles 'escaping to infinity' into believing that particles will not have effects in the vicinity of the hole. As Penrose points out

> . . . it seems to me to be quite unreasonable to suppose that the physics in a comparatively local region of spacetime should really "care" whether a light ray (or particle) setting out from a singularity should ultimately escape to "infinity" or not. To put things another way, some observer (timelike worldline) might intercept the light ray (or particle) . . . though he be not actually situated at infinity (and no actual observer would be so situated in any case). (1979: 618)

Consequently, the Hawking process of emission or evaporation from black holes may give rise to unpredictable behaviour throughout the causal future of each black hole. We should note that a definite asymmetry seems to have been introduced here. For an observer to the future of a black hole there is no uncertainty about events in its

†We have already noted that our knowledge of the interior of black holes is limited to energy/mass, angular momentum and charge. Any particles which manage to escape from a given hole will abide by this limitation and will possess the same total of these three quantities. Note also that Hawking is playing fast and loose with the concepts of position and velocity; as Nagel points out these are classical concepts, and it must be shown that they apply in quantum mechanics, not just taken for granted (1961: 305). Nevertheless, Hawking's main point here is that quantum laws are probabilitistic, not deterministic.

causal past. Because our knowledge of the likely physical properties of topological holes and naked singularities is necessarily and at best vague, I have taken the view that there will be uncertainty about their causal futures (for observers to their pasts) *and* their causal pasts (for observers to their futures). Of course, I agree that if we were fairly specific about the behaviour of a particular kind of naked singularity, we could reduce the uncertainty in, say, the causal past of the singularity—but a good deal would have to be said about the nature of any such singularity to enable us to do this. However, we have seen that black holes have fairly definite physical attributes, even if we cannot always pin them down precisely. And any imprecision will *always* concern the causal future of the hole.

Apart from the difficulty of determining events in the causal futures of black holes, there is a second related problem arising from the Hawking process. Given that black holes emit energy and are likely to do so even if they 'swallow' less and less matter, we can ask the question: do black holes ever 'disappear'? This problem is as yet unresolved, but Hawking believes that they might produce at the final moment a naked singularity; would this violate the cosmic censorship hypothesis? The answer to this depends on whether we take the hypothesis to apply only to gravitational phenomena *qua* GTR. Remembering our discussion concerning the geodesic incompleteness of space–times produced by naked singularities, it should be clear that they are embarrassments to GTR whether or not they are accompanied by quantum effects. The incompleteness of curves leads directly to indeterminacy, except in the cases of the initial and final singularities. Now D14 concerns the development of naked singularities on the basis of data sets using the field equations in the manner described above (p.144). We do not use any quantum mechanics. But in the Hawking process we treat the emission of particles semiclassically, using both GTR and quantum mechanics. Hence we can argue, as Hawking is said to have done, that cosmic censorship is 'transcended' rather than violated (see Penrose (1979: 617)). Hawking's argument seems to rely on two assumptions:

(i) that any uncertainty due to a quantum process is somehow less worrying than any due to simple geodesic incompleteness;
(ii) that we can make a fairly clear division between classical and semiclassical GTR.

Certainly, if the quantum Hawking process does occur in black holes, as seems likely, then we will have a shadow region in the causal future of the holes; but the shadow region here will not be quite so dark as the shadow regions of naked singularities and topological holes. At

least we can determine that any particle emitted by a black hole will very probably be in the form of thermal radiation and that it will have a well defined total of mass–energy plus angular momentum plus electric charge. Consequently we may decide that the unpredictability associated with the Hawking process is no more reprehensible that that arising from the behaviour of more typical thermal sources. On the other hand, we would be at a complete loss if asked to say exactly what might emerge from a naked singularity. Hence, there does seem to be some support for assumption (i). But if the quantum process leads to a naked singularity then it seems that we are in no better a position than if the naked singularity is a purely GTR phenomenon. In both cases we lose the ability to determine what might emerge from the singularity (when we are to its past) and what might enter it (when we are to its future). So although I agree that (i) is correct, it would not apply to naked singularities unless we could show that those singularities which arise from quantum physics are better behaved causally than from those which arise from GTR alone. And this—so far as Penrose's report shows—has not been done.

On the face of it assumption (ii) seems to be quite acceptable: there seems to be a clear difference between using GTR without quantum mechanics and applying quantum physics to the matter field in the vicinity of a black hole. However, if we accept that black holes are an indispensable part of GTR—since they are a consequence of the theory it would be strange to deny this—then we should count those processes involving black holes as GTR phenomena. Of course, if we use quantum physics to clarify the nature of such phenomena then we are modifying GTR. But if the Hawking process is involved in black hole physics this modification is essential. Consequently, GTR, when understood in this sense, would be in line with cosmic censorship only if *all* badly behaved naked singularities are ruled out. Fortunately, Hawking's proposal that a black hole might disappear leaving, momentarily, a naked singularity is very tentative indeed (see Hawking and Israel (1979: 20)), so we may still be able to adopt cosmic censorship and accommodate black holes in GTR space–times. But if Hawking's intuitions are right then I believe that cosmic censorship would be violated, at least with the hypothesis in its present form. It is also difficult to see how the hypothesis could be amended to take account of the occurrence of naked singularities in differing circumstances without making it appear *ad hoc*.

We shall now introduce our final group of concepts, each of which is associated with cosmic censorship. When a space–time is in line with the cosmic censor's wishes it is said to be *globally hyperbolic*; and a globally hyperbolic space–time is said to admit a *Cauchy hypersurface* (see Penrose (1979: 625)). A Cauchy hypersurface is an achronal slice

such that *all* events in a space–time may be determined by the data on the slice. We shall define it thus:

D15 For space–time ***M*** and achronal slice S in ***M***, S is a Cauchy hypersurface for the space–time if every point of ***M*** is in the domain of dependence of S; such a space–time is globally hyperbolic.

This definition is based on Geroch and Horowitz (1979: 251–2) and Penrose (1979: 624–5). We should note that not every hypersurface is a Cauchy hypersurface. For example, the plane $t = 0$ in Minkowski space–time is a Cauchy hypersurface, but the hyperboloid is not. The domain of dependence of the surfaces of simultaneity defined by the hyperboloid is restricted to events within the past and future light cones of the origin of Minkowski space–time, but there is no such limitation on the domain of the plane (see Hawking and Ellis (1973: 120, 205)). Quite clearly, if there are no naked singularities which might interfere with the causal determination of events from the data on a Cauchy hypersurface in a given causally stable, hole-free space–time, then that space–time satisfies the cosmic censorship hypothesis. Moreover, the fact that the Cauchy hypersurface enables us to determine all events in a space–time has suggested to many that Cauchy hypersurfaces may be linked with the idea of classical or Laplacian determinism, or simply *determinism* (see, for example, Malament (1977: 73)).

There are many difficulties involved in giving a precise definition to the idea of determinism; Nagel (1961: 278–85) and Earman (1971) describe the problems of trying to provide a thesis of physical determinism, and Davidson (1980: Essay 11) resists the extension of any physical thesis to include mental events. Nevertheless, Newtonian mechanics has long been regarded as the paradigm of deterministic theories; in the context of this theory Nagel gives us an intuitive idea of what we might expect from a deterministic theory:

> The position and momentum of a point-mass at a given time are said to constitute the "mechanical state" of that point-mass at that time, and (given a solution of the equations of motion for a physical system) . . . the mechanical state of the system at any time is completely and uniquely determined by the mechanical state at some arbitrary initial time. (1961: 279)

Because the equations of motion are symmetrical in time, there is every reason to suppose that events to the past as well as to the future may be determined in the same way. We must therefore reject the advice of Anscombe who says that we should only be concerned with pre-determination:

> . . . "each stage of the ball's path is determined" must mean "upon any impact, there is only one path possible for the ball up to the next impact" . . . to give content to the idea of something's being determined, we have to have a set of possibilities (provided by the theoretical context), which something narrows down to one—before the event. (1975: 73–4)

Of course, scientists are generally interested in prediction rather than retrodiction, but this does not mean that the determination of past events cannot be useful in certain circumstances—especially when we want to know exactly how a given system came to be as it was at a given time. Indeed, in the case of GTR, the analysis of events in past as well as future domains of dependence may prove invaluable in helping us to specify the properties of singularities.

The fact that events in GTR space–times *can* be determined should not persuade us to believe that GTR, like Newtonian mechanics, may be regarded as a deterministic theory. Bohm has argued that despite the complexities involved in the determination of events in field theories like GTR

> The theory is deterministic, at least in principle: the future and past motion of particles and the distribution of the gravitational field are predictable from the equations, if the situation at a given time is known . . .. (1964: 30)

Whilst there may have been some excuse for taking Bohm's point of view in 1964 when he wrote the above and before the work we have been considering in this chapter was done, there is little in favour of saying that GTR is deterministic without some very extensive qualifications. Certainly, we cannot rely on the rather simplistic idea of 'the situation at a given time'; as we have seen, we can only make the determination of events precise in GTR if we restrict ourselves to achronal slices in causally stable space–times and, when we are interested in the determination of events throughout a space–time, to the achronal Cauchy hypersurfaces. We can only expect determination to be successful in the regions designated as domains of dependence if those regions are free of topological holes and naked singularities.

Given a causally stable space–time, and an achronal slice in that space–time, we have seen that four facts must be taken into account when we try to determine events on the basis of the IDs for the slice.

(1) Determination may be prevented in the 'shadow' regions of topological holes and naked singularities.

(2) The likely quantum effects close to initial and final singularities may prevent determination.

(3) Probabilities only may be predicted in the causal future of black holes given the Hawking process of emission from the singularities.

(4) Events may be determined in regions free of all physical singularities and topological holes.

Consequently, we can only talk of GTR as being deterministic so long as we rule out naked singularities and topological holes and also ignore all quantum effects whether these are associated with the Hawking process of emission from black holes or with the regions of high curvature close to initial and other singularities. Any quantum phenomena which we feel obliged to consider within a gravitational context will be governed by quantum laws either in conjunction with the field equations, as in the case of the Hawking process, or by themselves, as may well prove necessary in the cases of initial and other singularities. The laws of quantum mechanics are, of course, indeterministic in the sense that with them we cannot determine events uniquely and completely: we must resort to the assignment of probabilities and statistical methods (see, amongst the many studies of this subject, Nagel (1961: 306–9) and Putnam (1975: vol. 1, essay 7)). We would also expect the laws of quantum gravity to be indeterministic in the same sense.

The field equations, and laws of motion, of GTR are often said to be deterministic (see, for example, Sklar (1974: 308)). We only need to consider (1)–(4) above for a moment to see that a theory with deterministic laws is not necessarily deterministic itself. First, there may be processes in the domain of the theory which require indeterministic laws, e.g. the Hawking process. Secondly, there may be phenomena which no law seems able to deal with and which violate determinism, e.g. topological holes. In cases (1)–(3) the non-quantum laws of GTR are deterministic laws in an indeterministic context. And in (4) the laws are deterministic in a deterministic context. The laws are deterministic in the sense that they only deal in probabilities of zero and one, and the context is deterministic in the sense that it in no way requires the use of indeterministic laws nor does it admit phenomena which violate determinism in the manner of topological holes. So only when the context is deterministic will we be able to ensure that we can take advantage of the deterministic features of the laws we use. Obviously, that context will be shaped by empirical and theoretical considerations, e.g. for GTR the physical reasonableness of topological holes or the nature of the properties of black holes.

Do space–times with Cauchy hypersurfaces present us with a deterministic context? The fact that they are in line with the cosmic censorship hypothesis enables us to rule out all naked singularities. We will of course need to make the physically reasonable assumption

that the 'Cauchy' space–times in which there are Cauchy hypersurfaces are hole-free. But what do we say about quantum processes and phenomena? For we know that if we allow these in a given space–time then they will interfere with the determination of events. If the Hawking process is admitted into GTR then we could not determine events in the causal future of a black hole which lies to our future. Hence a space–time with a black hole with Hawking emission could not have a Cauchy hypersurface since by definition every point of the space–time should be in the domain of dependence of the hypersurface. When a black hole with Hawking emission lies to the future of a hypersurface which purports to be Cauchy, the events in the causal future of the hole would not be in the domain of dependence of the hypersurface (which is therefore not Cauchy). Similar problems would arise if we admitted the quantum effects associated with the high curvatures close to initial and final singularities. A space–time with these effects could not have a Cauchy hypersurface since events close to the singularities would not be determined by the data on a hypersurface. Hence the minimum condition for Cauchy hypersurfaces is that only classical gravitational phenomena can be admitted into Cauchy space–times. Given this condition, it is clear that Cauchy hypersurfaces are deterministic.

We have defined naked singularities in terms of the determination of events; see D14 above. The cosmic censorship hypothesis, as we have interpreted it, excludes naked singularities and topological holes from GTR space–times. But we explained cosmic censorship in terms of the determination of events. We have discovered that black holes and intial and final singularities may all interfere with causal determination if the quantum effects associated with these phenomena are admitted into GTR space–times. So if the cosmic censor wishes to preserve the complete determination of events in GTR space–times, he must advocate a *stronger* version of his hypothesis. That is, he must say something more than the weak:

(i) GTR space–times should involve neither naked singularities nor topological holes;

he should demand that:

(ii) GTR space–times must not involve any effects, quantum or otherwise, which are likely to interfere with causal determination unless these effects are *totally* enclosed by event horizons.

Hence, the Cauchy hypersurface is in accord, not with the weak hypothesis (i), but with the strong cosmic censorship hypothesis (ii).

But hypothesis (i) is deterministic in nature and strictly demarcates the 'classical' solutions of GTR.†

The General Theory admits both deterministic and indeterministic laws, and both deterministic and indeterministic contexts. Should we be prescriptive and say that GTR is really a deterministic theory, and that any hint of indeterminism should be quashed? Should we prefer to see GTR as a 'classical' deterministic theory and thereby give the Cauchy hypersurface a pre-eminent status, rather than accept the idea that GTR should be viewed as a 'semiclassical' theory in the sense that at least some of its physically respectable models involve the use of quantum laws? We have already seen that semiclassical modifications to GTR may be essential if the theory is to remain empirically viable: we can hardly regard hypothesis (ii) as applying to the real world. However, there is little reason as yet to impose such modifications upon every GTR space–time.

If black holes do exist, and if they do emit particles, then we must account for their behaviour. We might resign ourselves to the quest for a new theory; but we might prefer to accept modifications to GTR which permit indeterministic as well as deterministic space–time solutions. If we follow the latter course, as I believe physicists like Hawking do, then we grant that Cauchy hypersurfaces do have an important role in GTR, but their determinism must be seen in a wider context. Perhaps the most useful space–times with Cauchy hypersurfaces are those defined by the Friedmann solutions. These space–times are a focal point for much work in modern cosmology. But even Friedmann solutions have their limitations. Whilst they give us good approximations to the behaviour of the actual universe back to the first 100 seconds, before this time the gulf widens between predicted behaviour and empirical evidence (see Hawking and Israel (1979: 4, 14) and Davies (1980: 159)). For instance, the radial motion of the matter contents of Friedmann space–times away from the initial singularity is always uniform—the solutions are homogeneous through and through; but in the actual universe it would be difficult to understand how galaxy clusters could have formed without large density fluctuations at early times (see Sciama (1971: 126)). These fluctuations could well be the result of quantum effects close to the initial singularity, which the Friedmann solutions (concerned, as they are, solely with classical gravitational phenomena) completely ignore. So we may say of Friedmann space–times that they apply in certain circumstances but not in all circumstances, and especially not when

†Although Cauchy surfaces have a role in quantum theory in determining the evolution of probabilities, we are concerned here with precise causal determination arising from the use of such surfaces in a 'classical' theory such as GTR.

we are faced with inhomogeneities on a large scale. Hence the Cauchy hypersurface of a Friedmann space–time must be seen from this perspective. Such a hypersurface allows us to explain a great deal about the gravitational behaviour of space–time, but there is a strong suspicion that its value will turn out to be limited by empirical considerations: by the possibility of quantum effects close to initial and final singularities, by the likelihood of Hawking emission from black holes, and by the inability of solutions like the Friedmann models to account for local irregularities. Consequently, to say that we should only allow space–times with Cauchy hypersurfaces in GTR would be presumptuous. And it is perhaps an overstatement to assert, as Graves does, that it is always reasonable

> . . . to expect that we should be able to break up spacetime into a family of wholly spacelike three-dimensional hypersurfaces . . . [with the result that] Einstein's theory is deterministic in just the same way as Newton's. (1971: 241)

The belief that GTR is deterministic can only be maintained with a defiant and questionable attitude towards the class of 'quantum' gravitational phenomena which we have met in this section. Graves and others who share his belief may take this position because they regard GTR as the pre-eminent classical theory and consequently they may be reluctant to 'dilute' the purity of GTR with quantum imperfections. Certainly any theory which admits quantum effects will not be deterministic. The fact that GTR may have to be seen as a semiclassical theory is perhaps, because of GTR's pre-eminence, more of a challenge to those who feel nostalgia for the deterministic dreams of Laplace. Nevertheless, the deterministic solutions of GTR have an immense instrumental value: despite the fact that the empirical standing of Cauchy hypersurfaces may be not as strong as some would hope, the group of space–times which possess such surfaces does have a considerable theoretical status. These space–times, which conform to the strong cosmic censorship hypothesis, do facilitate the understanding of the causal behaviour of 'normal' situations in GTR.

4.2 The cosmic censor

Having examined the story which physicists tell about the causal structure of GTR space–times, we are now in a position to give an overview of their tale. There are, perhaps, six different constraints which physicists may try to impose on GTR space–times. Each constraint is connected with causal behaviour, i.e. with how signals may propagate in space–time. They are listed below.

(1) Constraints on the differentiable structure.

(2) The assignment of a local time sense; i.e. the demand that space–time be temporally orientable.

(3) The local causality condition, which matches temporal and causal order and demands that the null signal is the 'fastest' signal.

(4) The condition of stable causality, which rules out closed time-like curves and other such anomalous behaviour, and which assigns a, not necessarily unique, global time sense to GTR space–times.

(5) The weak cosmic censorship hypothesis, as advocated by Penrose and others, which seeks to exclude only topological holes and naked singularities from GTR space–times.

(6) The strong cosmic censorship hypothesis, which demands that all quantum effects be excluded from GTR space–times, e.g. the Hawking process of black hole evaporation or radiation.

At each stage, we have noted that the decision whether or not to impose a constraint on GTR space–time is motivated by empirical considerations for the most part. For example, since closed time-like curves have no immediate empirical foundation, most physicists have no worries about adopting the condition of stable causality. But we must note that this does not mean that these physicists ignore space–times which are causally unstable. The properties of such space–times may bring into sharper focus those of better behaved space–times. The attitude of Hawking and Ellis may be summed up as follows: *if* there were closed time-like curves in GTR space–times *then* the following would be true about these space–times . . .. This should help us realise why causally stable space–times are preferable on a theoretical level—they are tidier, and they allow us to make predictions on the basis of initial data sets. As a matter of fact, there is no empirical evidence for these causal anomalies, so it is reasonable to prefer causally stable space–times on an empirical level also.

Of course, much the same approach may be found in some or other physicist's work in all the above demands for good behaviour. But as we seek to impose more and more causal constraints on GTR, more and more objections are raised. We saw that Clarke (1977) is a fairly radical opponent of the demand that GTR space–times should be well behaved causally. Where others are antagonistic towards closed time-like curves and naked singularities, he welcomes these 'disruptive' phenomena into GTR space–times. As far as he is concerned, these and other phenomena have not only theoretical standing in GTR (to sharpen our thoughts), but an empirical status which prevents us from ruling them out. His argument leaves much to be desired: for example, in the case of naked singularities he seems to ask us to accept

them wholeheartedly simply because their theoretical status is now so high that *some* physicists have come to believe that they actually must exist. But he forgets that what might be true of certain solutions of GTR, may not be true of the world.† And this takes us back to a general feature of the way physicists think about GTR space–times and their properties.

In Chapter 3, we mentioned that there is a tendency for physicists like Trautman (1967) to regard GTR as an abstract, mathematical theory. They often regard the empirical side of GTR as, at best, of secondary importance and, occasionally, as inessential to a proper understanding of the theory. There is opposition to this tendency, notably by Bondi (1967) and Sciama (1961, 1969): they argue that GTR is an *empirical* theory first and foremost.

If we believe that Bondi and Sciama are correct, as I do, then we must distrust Clarke's readiness to embrace naked singularities and other such physically implausible entities. Rather than say that these are the solutions of GTR and these are their properties, which we must take seriously, *qua* empirical status, simply because they arise in GTR solutions, we should say we shall take those properties of GTR space–times seriously, *qua* empirical status, if and when they deserve to be so taken. Hawking and Ellis (1973), Hawking and Israel (1979), Geroch (1979) and Penrose (1979) all maintain that the empirical status of closed time-like curves and naked singularities is very low indeed. Although a liking for abstract theories and an unhealthy disregard for matters empirical might persuade some physicists to credit such phenomena with existence in the physical world by virtue of their appearance in GTR solutions, we should not follow their lead.

Each constraint may be seen in two ways. First, as one which is demanded by the theory itself and perhaps without which the theory could not be made to work, i.e. give solutions etc, and secondly as one which is demanded by empirical evidence, or, at least, one which is made desirable by the available evidence. If we remove the stipulation that GTR space–times should be continuously differentiable manifolds, then the theory could not work—there is no ready-built 'discrete' geometry or other theoretical superstructure which would allow us to make anything near the variety of solutions and predictions which the above stipulation permits. The removal of the condition of time-orientability is almost as serious: the technical difficulties involved in constructing non-time-orientable space–times are tremendous. The other constrains—local causality, stable causality, weak cosmic censorship and strong cosmic censorship—are not so theoretically indispensable, and if we drop strong cosmic censorship the theoretical

†For examples of such solutions see Hawking and Ellis (1973: 161), Yodzis *et al* (1973), Eardley and Smarr (1978), Ellis and Schmidt (1977) and Clarke (1978, 1982).

loss for 'classical' GTR is nil, for this hypothesis merely rules out quantum effects in GTR space–times.

But when we ask about the *status* of such constraints in GTR we should not be concerned solely with theoretical status, but we should also direct our attention at the empirical standing of these constraints. If we do this then we get rather mixed results, compared to the rather strict ordering of theoretical status which follows the order of the contraints as listed—most disruption being caused by the removal of constraint (1) and least by (6). As Penrose (1974) points out, the empirical standing of the idea of continuous as opposed to discrete space–time is not so high as might be thought. No one has any serious objection to temporal orientability. But the idea of local causality is threatened by discoveries in quantum mechanics† and by 'backwards' causation interpretations of certain physical phenomena. Stable causality, leaving aside the agitation of Clarke and a few others, seems fairly secure. Weak cosmic censorship, which rules out topological holes and naked singularities, seems secure also. However, strong cosmic censorship is on very shaky ground, since it seeks to rule out effects which are linked to very plausible singularities. So, whereas we can draw a tidy picture of the theoretical standing of constraints (1) to (6), the view we have of the empirical standing of these constraints is ragged and sometimes uncertain. And this is exactly what we would expect, given the paucity of empirical evidence for the various theoretical concepts of GTR.

Therefore, whilst we have every theoretical reason to 'follow the crowd' and impose constraint after constraint upon GTR space–times, giving ourselves, finally, the predictive advantage of space–times in which all events can be uniquely determined from initial data sets, we have very few empirical reasons to do so. I do not advocate the 'chaos' of Clarke, but I do not subscribe to the 'order' of Penrose. The only reasonable verdict is that no clear-cut decision can be taken concerning the status of every constraint we have mentioned above.

There is, I must confess, an important assumption which underlies my verdict, namely that GTR is not a totally self-sufficient theory, but one that feeds off the theories and evidence of physics as a whole. Lucas, in his *Treatise on Time and Space* (1973) says that

> . . . in the General Theory of Relativity we become monists . . . it is an austere, all-embracing system, which views the whole universe as a unified self-subsistent substance . . . (1973: 241)

The substance Lucas refers to is space–time, which, he says, 'describes, accounts for and explains all phenomena' (1973: 241). This view is as extraordinary as it is popular. And it seems to be quite

†See Redhead (1981) on the Bell inequality experiments.

false. The General Theory does *not* describe and explain *all* phenomena; for, as we have seen, the apparatus of GTR is quite unable to deal with effects close to singularities, even if we allow the use of quantum GTR methods. The General Theory predicts effects which it *cannot* fully explain. At best we are given an indication of the direction in which the answer lies. It is a singular mistake to see GTR as a self-subsistent enterprise. A glance at any standard text-book on GTR will reveal the intimate relationship GTR has with STR and electromagnetism. If we take a view like Friedman (1983), the very 'substance' proposed by GTR is *exactly* that proposed by electromagnetism and STR, namely space–time; and all motions in the manifold, whether gravitational, electromagnetic or kinematic may be referred to this entity.† Many of the constraints under discussion in this chapter gain or lose their plausibility from the evidence, not just of GTR, but of other theories also. Furthermore, recent texts on GTR show the developing relationship between GTR and quantum mechanics, so much so that certain gravitational phenomena may only be 'explained' by use of the methods of both theories (see Hawking and Israel (1979) and Zel'dovich (1979)). If we become monists when faced by GTR, we run the distinct danger of losing touch quite rapidly with physical reality, for GTR is quite clearly informed by many of the rest of today's physical theories.

My final confession is that my remarks above display a Quinean devotion to the interplay of our scientific ideas. Quine's famous claim, that

> our statements about the external world face the tribunal of sense experience not individually but only as a corporate body. (1953: 41)

seems to be vindicated, at least in the context of modern physics.‡ There is undoubtedly a vast array of interconnections between these theories, and assertions in one will depend on varying ways on assertions in others. I have pointed to just a few ways in which GTR links up with other theories; but I am convinced that many more would be revealed by an investigation more thorough than mine. Of course, implicit in what I have said is the claim that an assertion in one theory, say 'classical' GTR, could be overturned by assertions or developments in another theory, say quantum mechanics. This claim seems to be justified by my discussion of the Hawking process of black hole radiation. In 'classical' GTR, no radiation is predicted. This leads

†Whether this entity may be reduced to material interactions is, of course, another matter.

‡So long as we are not too sensationalistic in our idea of sense experience.

to the embarrassing consequence that black holes will get larger and larger. But quantum methods show us how particles can leap out of apparently impenetrable barriers. Hence, the prediction of 'quantum' GTR is that there will be radiation. So assertions in one area of physics may overturn assertions in another. This sort of reasoning fits quite neatly, of course, into the 'network' model of theories advocated by Duhem, Quine and Hesse (see Hesse (1974)).

In the next chapter, I shall be concerned to demonstrate the fluidity of GTR, both in the sense described above *qua* its relationship to the rest of physics, and in the sense that GTR is not a fixed, static, monolithic theory but one which is dynamic with uncertain boundaries, and with, at times, uncertainty at its heart. In doing so, I shall be challenging the approach of those who, like Kuhn and Lakatos, advocate a view of scientific theories in which they are, for certain periods, immune—at least in their allegedly essential details—from revision. I believe this challenge to be a direct consequence of a strict adherence to a thorough-going network model of scientific theories: revisions may occur anywhere and at any time.

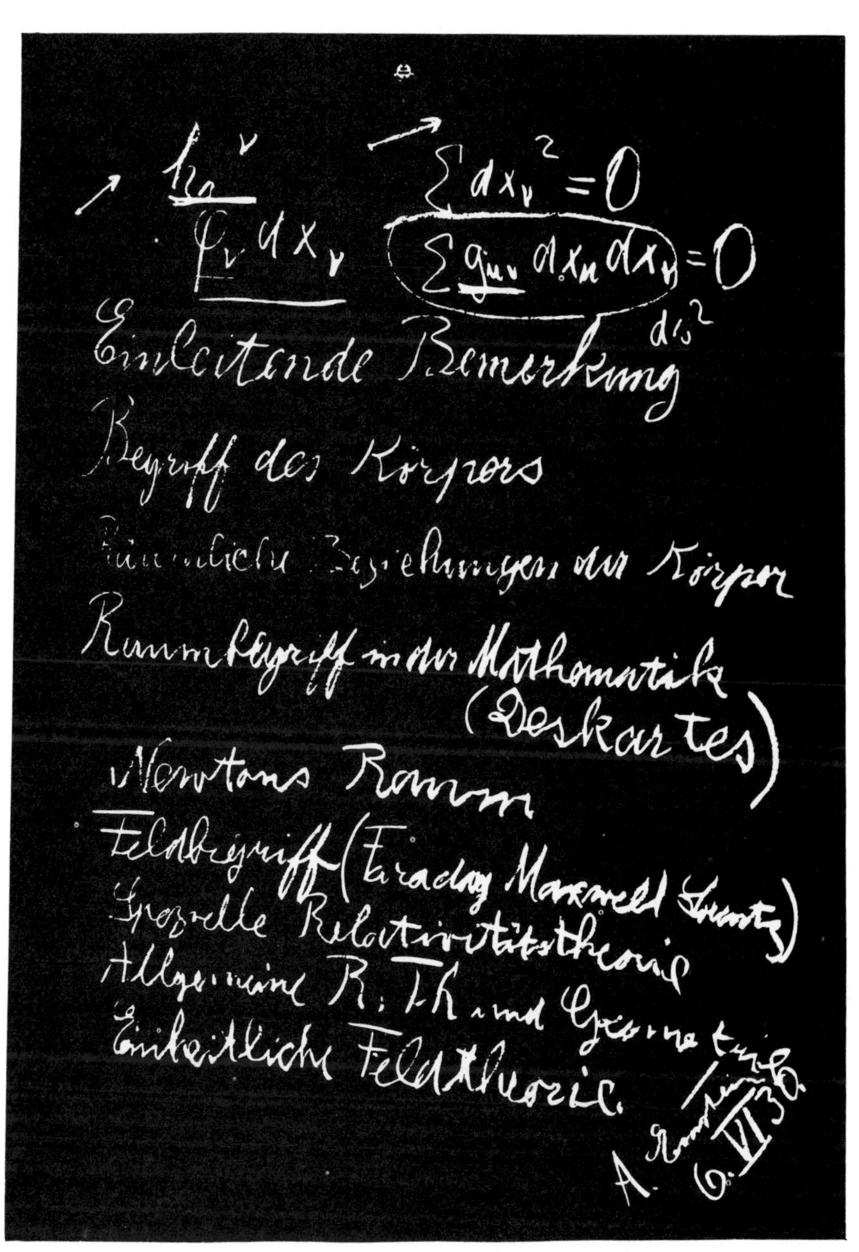

The blackboard used and signed by Einstein during a lecture at Nottingham University on 6 June 1930. Courtesy Dr M Heath, Physics Department, Nottingham University.

Chapter 5

Relativity—Dead or Alive?

Introduction

What is GTR? Is it the relatively neat and tidy theory sometimes portrayed in elementary text-books? Or is it more like a patchwork quilt, incorporating many, possibly conflicting, beliefs? Is it a stable organism, which gives comfort to those who claim to have identified its essential features? Or is it a *dynamic* creature, full of life and change? Are the field equations *the* focal point of the theory, or should we allow other elements of the theory to have a dominant role? What are the field equations of the theory—the Einstein equations, with or without the cosmological constant, or Raine's integral equations? Can we answer questions about the status of Mach's principle *now*, or do we have to wait for the death of GTR?

In this chapter I shall explore various issues concerning the development, structure and contents of GTR. These issues will influence our response to all the above questions. I shall argue that the most plausible view of GTR is that it is an evolving theoretical context, and that without such an evolution gravitational research would be impoverished. But any dynamic situation needs controlling features. I shall claim that these derive from the scientific community's search for simplicity in the structure and content of their discipline. A theoretical context is held together by a network of ideas which provides maximal coherence both within that context and with the key features of other relevant scientific contexts. Any theoretical discipline must therefore change as its internal features evolve and as physical science as a whole moves on. The study of GTR provides us with a crucial piece of evidence for the view that all scientific theories evolve dynamically. Static views may have instrumental value for the historian or philosopher intent on presenting a particular characterisation of a theory, say in the form of a time-slice of GTR in 1929, but we should resist the temptation to place too many strictures on the

development of science. The 'descriptive' accounts of the history of science by Kuhn, Lakatos and others have taught us much about the structure of theories. But in their eagerness to provide comprehensive descriptions they have over-simplified the actual structure of scientific theories: it is far less coherent than they maintain. I will argue that this is certainly so in the case of GTR, and, if Pais (1986) is correct in his discussion of particle physics, the rigidity offered by Kuhn and others is also inappropriate when trying to characterise other domains of science.

5.1 Debate and dissension in GTR

We have observed a number of issues which have given cause for debate and dissension within the context of GTR. In this section, we shall review these and discover several other related issues.

5.1(a) Topological and causal constraints

How far should these go? And upon what criteria, if any, can we base a distinction between the plausible constraints and the implausible? There are, perhaps, five main areas of concern.

(1) Concerning the differentiable structure; are we right in assuming that the structure of space–time is such that it is continuously differentiable? We might take the view that it is discrete, as may be the case in a quantised space–time theory in which the basic spatio-temporal background is not composed of points, but, say, of twistors as suggested by Penrose (1974) and others. But, here, one might wish to say that the empirical acceptability of such a quantum theory of gravity would be a decisive factor. Other competing quantum theories, e.g. those based on the path integral approach, may prove to be more plausible empirically. At the moment it is very difficult indeed to make any decision about the ultimate spatio-temporal structure which will emerge in an acceptable quantum theory of gravity. But it seems difficult to deny that the empirical evidence will favour one rather than the other of the two possible structures—discrete and continuous. Nevertheless, we might question the use of differential rather than difference equations in dynamics. Use of the former implies faith in the supposition that the spatio-temporal structure is based on the real rather than the rational numbers. As Newton-Smith (1978) points out, there seems to be no empirical evidence which could settle this issue. We may try to judge between these two possibilities on grounds of simplicity, as Hesse (1980) suggests, but the claim that simplicity is a sign of the truth is far from uncontentious.

(2) Concerning time-orientability and precedence; as we saw, the assignment of a local time sense to all points of the manifold, and the ordering of all non-space-like connected points with temporal relation of precedence, seem to be basic and empirically indisputable features of the space–time structure. Of course, one could conjure up solutions which were not time-orientable; but these would be ruled out by all as empirically unacceptable.

(3) Concerning local causality; the local causality conditions assume that causal order is no different from temporal order and that no causal signal can travel faster than the null signal. Here we meet with rather more contention. The latter assumption is clearly an empirical matter; for example, the Bell inequality experiments in quantum mechanics may show conclusively that in certain circumstances causal signals can travel faster than light. As far as the former assumption is concerned, we must meet with the debate concerning backwards causation—causation in the reverse temporal direction. Whether we can rule out backwards causation on *a priori* or *a posteriori* grounds is a difficult problem, and although I am inclined to think that the introduction of human agency into the debate tends to favour a non-empirical resolution to the problem, I am also inclined to keep an open mind and allow that there may be some empirical evidence which involves the human factor and which suggests that backwards causation can take place. Of course, this may rule out free will as a concept which can be applied to *all* circumstances in which humans find themselves. Without human agency, I think backwards causation may be a distinct possibility—but what grounds do we have for divorcing our scientific theories from human activity? But, we must acknowledge that only in the Hawking process of emission from black holes does this issue arise in GTR. And there, whether or not backwards causation occurs is a matter of interpretation—we can quite easily say that it does not with no significant theoretical loss.

(4) Concerning causal 'anomalies', e.g. closed causal curves; the condition of causal stability assigns a global time sense to space–times allowing us to rule out closed or almost closed time-like curves and closed null curves. Closed curves allow the possibility of a signal from a point reaching some other arbitrarily close point in the backwards light cone of the point of origin, and in fact returning to the origin. Although this is not backwards causation proper, it carries with it the same discomforting consequence, namely the fact that an event can be a cause of an event to its past. Such possibilities do arise in certain GTR space–times, e.g. the Gödel solution, but, again, we may be forced to say either that human agency allows us to rule out such causal anomalies or that human agency may not always be free. Fortunately, there is no hard empirical evidence to support the idea of closed causal curves etc.

(5) Concerning further anomalous causal behaviour in space–times, e.g. naked singularities; the cosmic censorship hypothesis is designed to isolate well behaved classical solutions of GTR (see Penrose (1979)). I have presented two versions of the hypothesis: strong and weak. The weak hypothesis rules out naked singularities and topological holes; the strong hypothesis goes further, ruling out all phenomena likely to disrupt causal determination, e.g. the Hawking process of black hole evaporation or quantum effects in the early universe. Whilst there is empirical justification for the expulsion of topological holes from GTR space–times, there is less of a case for the exclusion of the other phenomena mentioned above. These are all gravitational phenomena which, most physicists would agree, cannot be ruled out as yet. Both hypotheses, if they are meant to have ontological significance, are therefore very bold conjectures. For both run against the empirical current; black holes, together with Hawking evaporation, which allows us to treat black holes as thermodynamic objects, are very respectable nowadays, and naked singularities are becoming less reprehensible as each year passes (see Davies (1980)). Only the weak hypothesis has any real chance of being regarded as one which GTR obeys. But, as I argued in Chapter 4, that hypothesis is far less coherent than the strong one, and has every appearance of being *ad hoc*. For the strong hypothesis rules out *all* causally disruptive behaviour, but the weak hypothesis is more selective, ruling out just those disruptive phenomena which are less likely to obtain in the actual universe. However, the strong hypothesis can hardly be true of the real (quantum) world. Of course, the weak cosmic censor can justify his hypothesis by claiming that it is concerned with *classical* GTR and not with quantum effects which might intrude into this tidy picture; but this justification rests upon the presupposition that the split between classical and quantum GTR has ontological significance. If it does not, as I believe, then all we can say of the weak cosmic censorship hypothesis is that it has an instrumental value, which lies in the fact that it allows us to isolate and to investigate the properties of a class of relatively well behaved space–times.

5.1(b) Quantum versus classical GTR

This debate arises from the foregoing considerations concerning the activities of the cosmic censor. As I say above, I do not think that a distinction can be drawn between quantum GTR and classical GTR which has anything more than instrumental value. Such a distinction arises from the tendency to see GTR as a self-contained mathematical theory of gravitation and to forget that its proper role is as an empirical theory, the purpose of which is to explain and describe all gravitational phenomena so far as it can. The subject matter of GTR is

gravitation, and it should try to deal with this regardless of any quantum effects which intrude into the gravitational context. There are, of course, circumstances in which the gravitational phenomena under consideration seem to lie outside of the domain of GTR, even when quantum methods are utilised, e.g. very close to a singularity. But with instances like the Hawking process, GTR together with quantum methods remains intact, and delivers what most believe to be a good first approximation to the result that a thorough-going quantum theory of gravity might give us. It is interesting to note that physicists do put a great amount of trust into the results of GTR plus quantum methods. Their assumption that such joint ventures will turn out to be successful is based on the individual successes of quantum mechanics and GTR. There is, of course, no *a priori* reason why the consideration of theories, which are successful in their own domains, should prove to be even approximately right in situations with which neither theory can deal by itself. Nevertheless, there are considerations which support such an enlargement of the context of GTR. Quantum methods have been used in conjunction with classical electromagnetic fields; there is, moreover, experimental confirmation of the prediction given by this approach. If we consider a hydrogen atom, we can obtain the hydrogen spectrum and energy levels from the wavefunction for the electron, which is arrived at by using a background classical electromagnetic field and treating the electron as a quantum particle on that field. This procedure is similar to those used in black hole physics and in quantum cosmology. The root assumption which underlies the quantum GTR approach is that the large-scale gravitational phenomena described in classical solutions of GTR are really 'built' from small-scale quantum effects. The relationship between quantum gravitational effects and classical effects is in some ways analogous to that between statistical mechanics and thermodynamics. In both these relationships the 'global' macroscopic picture is held to be the resultant of the various micro-effects. Again, the success of the thermodynamics–statistical mechanics approach encourages relativists who utilise quantum methods in gravitational contexts.

5.1(c) Constraints on the metric

There is one popularly accepted constraint: the metric of GTR space–times should be of Lorentz signature, as discussed in Chapter 2. But after that, when it comes to choosing which metrics to use in GTR solutions, almost anything goes. Of course, some metrics find much more acclaim than others. The Robertson–Walker metric is often considered to be the most appropriate for the description of the large-scale behaviour of the universe; this metric is the one employed

by the Friedmann models. Moreover, Raine argues that his Machian criteria, employed in conjunction with his amended field equations, single out Robertson–Walker space–times as Machian in character (see Raine (1975)). But, regardless of Raine's arguments, solutions with the Robertson–Walker metric are perhaps the paradigms of relationist solutions in GTR but only if we can rule out empty solutions with the Robertson–Walker metric (see O'Hanlon and Tupper (1972)). We should note that not only do the Friedmann models have this metric, but there are solutions of the Brans–Dicke equations and the Hoyle–Narlikar equations which have the Robertson–Walker metric also (see Raychaudhuri (1979: 168*f*, 200*f*)). However, many other metrics are employed in conjunction with the Einstein equations, and also the Brans–Dicke and Hoyle–Narlikar equations. Should we have constraints on the range of metrics permissible in GTR? We have seen that Hawking and Ellis (1973) suggest that we should exclude that Gödel metric on physical grounds, and the metrics of inexact solutions receive much the same treatment. We have also noted that Raine tries to rule out a class of metrics on mathematical grounds—but this act of exclusion depends on an acceptance of his integral form of the field equations.

5.1(d) The content of matter fields

Two fundamental constraints on the matter fields of GTR are, first, that the field equations obey the local law of energy and momentum conservation† and, secondly, as stipulated by Hawking and Ellis (1973), that local kinetic energy is never negative. Both of these constraints are supported by all available empirical evidence. Further demands upon the matter fields are perhaps less convincing. The relationist might wish to exclude solutions of the Einstein field equation where the matter tensor is zero. This cuts down on the generality of the Einstein field equations, but, as I have argued, the relationist can justify his move by maintaining that all physical laws have an implicit range of application and that, in the case of the field equations of GTR, perfectly empty space–times, like Minkowski space–times, fall outside this range. There are other moves which can be made to dispose of 'empty' solutions. The most notorious is Einstein's advocacy of the cosmological constant: if this constant is non-zero, then empty space–time is not a solution of the field equations. Einstein's motivation for the inclusion of the cosmological constant in his field equations centred on his desire not to rule out a static universe as a solution of the equations. With the constant at an appropriate level, the equations can deliver a model for a static

† See Rindler (1977: 180).

universe which is closed and spherical. As we have observed, events—both theoretical and empirical—soon overtook Einstein. On the one hand the development of the Friedmann models and of the general form of the Robertson–Walker metric gave theoretical respectability to dynamic, expanding/contracting solutions, and on the other hand Hubble and other astronomers discovered that there is a progressive shift in the spectral lines (towards the red) for almost all galaxies we can observe (the shift in wavelength with distance; see Hubble (1936)). This Hubble shift leads directly to the inference that the universe is expanding and is not, as Einstein supposed that it might well be, static. There is, of course, no need to believe that we are at the centre of this expansion: if we accept the cosmological principle, then we would expect that from the vantage-point of each galaxy in the universe, almost all other galaxies would be receding from the point of observation. The most popular analogy used to demonstrate this kind of expansion is that of dots on the surface of a balloon which recede from each other as the balloon is inflated—no one dot is regarded as the centre of the expansion. These theoretical and empirical advances undermined the use of the cosmological constant in solutions of the field equations, and for many years the constant was given little or no serious consideration by workers in the field. Indeed, Einstein is reported to have said that the inclusion of the constant was his greatest mistake (see Misner *et al* (1973: 410)). In recent years, however, there has been a revival of interest in the constant. Recent work in quantum cosmology indicates that the constant may be non-zero in certain quantum solutions of the field equations. Most cosmologists are reluctant to drop the constant altogether as it provides a far richer range of models which, some think, may be of more than theoretical interest (see the articles by Rees, Dolgov and Wilczek in Gibbons *et al* (1985), and also Hawking (1983)). Despite Einstein's feeling that the constant was a mistake, Misner *et al* point out that

> A mischievous genie, once let out of a bottle, is not easily re-confined. (1973: 411)

There are two other possible ways to attempt to exclude empty solutions. First, we can adopt the field equations of Hoyle and Narlikar, in which the gravitational constant can vary (see Narlikar (1977) and Raychaudhuri (1979)).† A consequence of these equations is that in empty space–times there will be no sense to talk of the inertia

† The Hoyle–Narlikar equations employ a 'C-field' which facilitates the *direct* interactions of distant particles which in turn aids the spontaneous creation or annihilation of particles. See Hoyle and Narlikar (1974) for technical details.

of a test particle, hence from a gravitational point of view such space–times will be physically meaningless. Secondly, we may follow Raine's approach and try to rule out empty solutions on mathematical grounds.

Another constraint on the matter field is suggested by Brans and Dicke (1961).† They argue that we may add a scalar field variable to the basic Einstein field equations. Their motivation is unashamedly Machian—the resultant scalar field is determined by the distribution of mass–energy in the space–time under consideration, providing the long-range gravitational effect which Mach had presumed might come from the distant fixed stars. A central feature of the Brans–Dicke formulation is that, as in the Hoyle–Narlikar field equations, the gravitational constant varies. A number of 'scalar' field variants of GTR, along the same lines as but more sophisticated than that of Brans and Dicke, have flourished in recent years (see Will (1979) who presents a survey of these). Although most physicists prefer Einstein's formulation of the field equations, thus resisting the idea of a scalar field determined by the structure and evolution of the universe, the empirical evidence is not decisive in favour of their preference. Cosmological tests have not been as decisive as some would have hoped (see Raychaudhuri (1979: 174) and Will (1979: 62, 88–9)). The empirical standing of the Brans–Dicke approach has been helped by the fact that the solutions of their equations include Robertson–Walker solutions, as is the case with Raine's approach. But rigorous tests in the solar system have proved far from satisfactory for the Brans–Dicke formulation. The equations of their theory contain a constant ω; as this constant increases in size, their equations begin to approximate to the Einstein equations and the scalar field has less and less effect.‡ Observations of the Viking orbiter of Mars suggest a large value for ω and therefore at best a minimal role for the scalar field. Although this might be regarded as evidence against the Brans–Dicke

†Brans and Dicke adopted an approach similar to that of Sciama who argued that inertial forces arise from a component of gravitational interaction which is inversely proportional to distance, with

$$GM/Rc^2 \sim 1$$

where G is the gravitational constant, M is the mass (finite) of the visible universe, and R is the radius of the boundary of the visible universe. Of course, if we accept that causal signals can travel no faster than light, then only the visible (from point p) universe can causally affect events (at point p). As a body accelerates, the inertial forces present are a measure of the reaction of the rest of the (visible) mass of the universe or the body (see Sciama (1953, 1961)).

‡ A similar situation arises in the BWN–Bekenstein scalar field approach which involves both ω and Λ. As $\omega \to \infty$ and with Λ close to zero, the approach collapses to GTR (see Will (1979: 48)).

approach, it cannot rule out the strategy of utilising scalar fields (see Adler *et al* (1975: 385)). But as ω increases, so the *ad hoc* character of the scalar field becomes more evident.

5.1(e) The field equations

The discussion above introduces us to two questions concerning the form and content of the field equations: should we use one form of field equation rather than another, and does the use of certain forms push us outside of the context of GTR? There are a good many alternatives around, each purporting to give us a satisfactory basis for the investigation and explanation of gravitational phenomena. Four of these are treated by many writers as being within the context of GTR.

(1) The Einstein field equations $\boldsymbol{G} = k\boldsymbol{T}$.

(2) The above equations with the addition of the cosmological constant.

(3) The Brans–Dicke type of field equations, which incorporate a variable scalar field together with those of (1).

(4) The integral form of the field equations advocated by Raine and others.

A feature of (3) and (4) which we noted in Chapter 3 is that, under certain conditions, these equations reduce to the 'usual' Einstein form. This is also true in the cases of (2)—when Λ is set at zero—and of the Hoyle–Narlikar equations. However, the Hoyle–Narlikar equations are rarely regarded as falling within the context of GTR: a fact which owes much to the instantaneous action-at-a-distance which is demanded by their theory. Furthermore, each of the above forms admits the Robertson–Walker metric as a solution; Raine's approach does this and more, giving this metric a fundamental status. Of course, the Brans–Dicke (and Hoyle–Narlikar equations) give us predictions significantly different in some respects from those of the Einstein equations, most importantly that the gravitational constant varies. The Raine approach also gives particular empirical significance to certain features of Robertson–Walker models, namely homogeneity and isotropy. Hence, with the right empirical evidence—provided it was sufficiently strong—we could make a decision about which form of the equations should be used to describe and explain gravitational phenomena. Even though we might be inclined to favour Raine's approach, since this accords well with the conclusions of most modern cosmologists, the twin assumptions of homogeneity and isotropy are coming increasingly under fire. On the one hand Ellis' model presents a theoretical challenge to these ingredients of the cosmological principle and on the other hand significant deviations from overall homogeneity have now been

observed as suspicions grow that the universe's expansion is not as smooth as the principle demands (see, for example, *Scientific American* editorial (1986 **254**(6) p. 16)).

So, if we concentrate our attention on empirical matters, taking our lead from cosmology, we will have difficulty in deciding between the above formulations, including between (1) and (2), given our remarks concerning the cosmological constant in the sub-sections above. But do these formulations belong to different theories, or to distinct variants of Einstein's theory, or to one single theoretical context? In Chapter 3, we described a model of GTR as essentially consisting of

(i) a differentiable manifold—the space–time;
(ii) a metric on the manifold of Lorentz signature;
(iii) a matter field which could be described by the stress-energy tensor;
(iv) a set of field equations relating (ii) and (iii).

Now we could be quite dogmatic and say that (iv) should be the *Einstein* field equations. But what reasons other than historical ones could we offer for this rigidity? And would we choose Einstein's equations with or without the cosmological constant? It is interesting to note the attitudes of physicists to the alternative formulations we have mentioned. Recent work done on singularities and the very early universe suggests that the field equations of GTR are a good first approximation, but require modification if we are to give a sound description of quantum effects (see Gibbons *et al* (1985)). There is almost no hesitation at all in allowing the cosmological constant into the theoretical context of GTR—only its empirical status is in doubt, despite the fact that this constant introduces an underlying sourceless field into solutions of the field equations. But when the question of a *variable*, scalar field is forced, the response is not so unequivocal. Brans and Dicke, in introducing their ideas, say they are presenting a relativistic theory of gravity, but one which is not in line with the equations of general relativity. Many writers refer to the 'Brans–Dicke theory' (see for example, Raychaudhuri (1979) and Will (1979)). The impression is given that, despite the extensive common ground shared by Brans and Dicke and Einstein, they must be treated as quite separate theories since they lead to different predictions. This also seems to be the verdict on the Hoyle–Narlikar approach. But, the introduction of a cosmological constant, at a level in line with current observational limits, also leads to different predictions, even if these are concerned only with large-scale, cosmological effects (see Rindler (1977: 235–8)). I am tempted to believe that those who are willing to embrace the cosmological constant within the context of GTR do so for the same reason that they tend to exclude the Brans–Dicke scalar

field, namely that Einstein himself added the constant, but not the variable field. Of course, one could argue that an additional constant does not imply a departure from the essentials of a theory, but that an additional variable does. This might follow from a commitment to Will's strongest version of the equivalence principle, SEP, which rules out terms which postulate gravitational fields over and above the metric g (see the discussion of equivalence in §2.1 and Will (1979, 1981)). In particular, SEP rules out the Brans–Dicke formulation and the Hoyle–Narlikar approach. But the inclusion of the cosmological constant does not by itself violate SEP. Three problems stand in the way of our accepting SEP as a reason for accommodating the cosmological constant but not scalar fields etc. First, Will admits that no formal proof exists which shows that the assumption of SEP yields only GTR. Secondly, we could regard the idea of minimum coupling as a methodological demand for simplicity and consequently we should resist the addition of any curvature term to the field equations. Thirdly, the scalar field and the cosmological constant both carry ontological implications along with them: in effect, both constants and variables say about the physical world that some state of affairs obtains, in the case of the cosmological constant that there is a sourceless field, and in the case of the scalar field variable that there is a variable field due to the action of distant masses.

We should note, however, that some authors *do* regard the Brans–Dicke equations as a *variant* of Einstein's GTR, and therefore not completely outside its theoretical context (see, for example, Adler *et al* (1975)). A comparison of the field equations of Einstein and Brans–Dicke demonstrates why one might be inclined to regard the latter as a variant of the former:

$$R_{ij} - \tfrac{1}{2}Rg_{ij} = kT_{ij} \qquad \text{(E)}$$

$$R_{ij} - \tfrac{1}{2}Rg_{ij} = \frac{k}{\phi}(T_{ij} + T^{\phi}_{ij}) \qquad \text{(BD)}$$

where the gravitational field is described in (E) by the metric g and the quantities derived from it, and in (BD) by g *and* a scalar field ϕ. The tensor T^{ϕ}_{ij} may be thought of as the energy momentum of the scalar field. † The similarity between (E) and (BD) is striking, and it is obvious that the sum total of Brans and Dicke's interference is the simple addition of a scalar field to the basic Einstein tensor fields. As argued above, no more ontological liberties seem to be taken than when the cosmological constant is added to the Einstein equation, for in this case we assert the existence of a *sourceless* field.

† The expression on the right-hand side of (BD) involves the dimensionless constant ω.

So far, we have considered changes in the *form* of the field equations which carry with them changes in the *content*. But what about Raine's formulation? Here no changes in content are proposed, but only a variation in form which is supported by mathematical arguments for its validity. I can see no reason for any strong objection to the claim that Raine's formulation belongs to the theoretical context of GTR. Of course, Raine's equations do not deliver the rich variety of solutions which Einstein's equations give us; but that is the point of his method. He believes that his approach singles out the most empirically acceptable solutions of GTR. A significant problem for Raine is the fact that any evidence for his approach is also evidence for GTR. There seems to be no straightforward way of isolating empirical evidence in favour of Raine's equations alone. Consequently, an advocate of Raine's approach must argue *for* the pre-eminence of global solutions of GTR and *against* local solutions on grounds of economy as well as of mathematical coherence. Consquently, the richness of GTR must be regarded as needless ontological extravagance.

5.1(f) Empirical data and cosmology

This is a rich area of conflict which underlies many of the issues we have already mentioned. In the course of Chapters 3 and 4 I highlighted four debates:

(i) concerning homogeneity and isotropy and the adoption of the cosmological principle;
(ii) concerning the cosmological constant;
(iii) concerning singularities;
(iv) concerning spatial closure.

Each of these debates falls largely within the domains of cosmology and quantum cosmology. One fact we have already noted is the lack of clear-cut empirical data which might conclusively settle these, and other, debates. Indeed, not only do the above matters remain unresolved, the empirical standing of GTR itself is not so firm as some would wish. We can build many cosmological models, the essential details of which may be compatible with more than one formulation of relativistic gravitational theory. As we have seen, solutions with the Robertson–Walker metric may be given not only for the Einstein field equations (with or without the cosmological constant), but also for Raine's formulation and for the Brans–Dicke equations, as well as for the Hoyle–Narlikar theory.

This is an interesting situation. Two main alternative procedures seem to present themselves.

(1) We could put our faith in Einstein's field equations *or* in Brans–Dicke *or* in some other formulation, and proceed to test their

predictions against the empirical evidence, the bulk of which will be drawn from modern cosmology.

(2) We could put our faith in those empirical features of modern cosmology which seem to be conclusive, e.g. redshift, approximate homogeneity and isotropy, properties of microwave background and so on, and proceed to assess the alternative theoretical views with these features in mind.

What separates these two procedures? In (1) *theory* comes first in the scientist's mind, and empirical evidence is an instrument which gives him the means to test his ideas. In (2) priority is given to *empirical evidence* by the scientist, and various theoretical viewpoints are assessed on the basis of the evidence. Note that I make no claim for *logical* priority of theory in (1) or empirical data in (2). Indeed, I believe that neither has such a priority. Undoubtedly scientists often give *methodological* priority to either theory or to evidence. Mach, of course, is an example of a scientist who gave not only methodological priority to observation and experiment, but logical priority also—we can have observation without theory, but theories require empirical data, since they merely consist in the economical *description* of the physical world. But those who accept the thesis that observation is theory-laden, as I do, must oppose Mach's gift of logical priority to empirical data. But, even if we accept the theory-laden observation thesis, there is no need to deny that observation and experiment might be given methodological priority. And it is the question of which is given this priority—theory or empirical evidence—which distinguishes (1) and (2) above. There are plenty of scientists and philosophers who prefer to give methodological priority to theory, and therefore to follow the path (1). We saw in Chapters 1 and 2 that Einstein himself, especially in his later years, thought that theoretical concerns should play a leading role in scientific investigations. But just as there are dangers involved in giving empirical data an absolute priority, we can also go overboard as far as theory is concerned: as we have observed, Trautman (1967) admits that he and many of his colleagues sometimes regard GTR as a purely mathematical enterprise, with theoretical consistency and simplicity uppermost in their minds, forgetting that what is said does have physical implications.

There is, however, a third possibility: we could combine (1) and (2). Thus we would admit the various formulations of the field equations into GTR's theoretical context *and* we would give empirical features of modern cosmology a central but not dominant role in that context. This has the advantage of not giving too much emphasis to empirical matters by themselves, especially when the evidence for GTR is hardly as good as that, say, for quantum mechanics, or to theoretical concerns by themselves, especially when these can lead to

an almost purely mathematical account of GTR. Indeed, the gift of a dominant role to empirical evidence has distinct dangers, since the evidence for the 'orthodox' view of cosmology is coming increasingly under fire (see, for example, Hart and Davis (1982) and Birch (1982) as well as Ellis (1978) and much of the discussion above).

5.2 What should we count as 'relativity'?

We could ignore much of the debate and many of the questions in the preceding section. We could adopt a fairly rigid attitude towards the structure and content of GTR. We could insist that

(i) the field equations of the theory are the Einstein field equations, of 1916–17, with no additional terms;
(ii) the local causality and energy conditions obtain for the matter field; and
(iii) quantum methods may not be involved in GTR.

Then, following some elementary text-book view, we could maintain that GTR may be split up, quite neatly, into the following components:

(i) a continuously differentiable four-dimensional manifold;
(ii) a metric of Lorentz signature;
(iii) a well behaved matter field which obeys the two basic mass–energy conditions above;
(iv) the *Einstein* field equations.

We could then say that only a model consistent with components (i)–(iv) is a model of GTR.

As we saw in Chapter 3, Sklar and many other absolutists$_1$ seem to view GTR in this conservative way. If we follow their lead we would then be able to maintain that the approaches of Brans and Dicke, of Hoyle and Narlikar, of Sciama and Raine, and of Hawking (in black hole physics) and Zel'dovich (in quantum cosmology) do not belong to GTR. We would have proposed a neat and tidy, 'classical' framework for GTR which have as its unshakable foundation the field equations proposed by Einstein.

But if we decide to pay attention to the issues raised in the last section, then we ought to consider modifying, perhaps dramatically, this narrow, 'classical' view of GTR. We would no longer be confined to a theoretical structure which, despite its important applications, has nevertheless a limited scope. We could then contemplate answering questions about a wider range of gravitational phenomena than the 'conservative' models of GTR encompass. For example, we could attempt to describe the problem of singularities in a GTR

context, for, after all, singularities are consequences of the theory even when conceived narrowly. But we should be careful not to deny the importance of many well established solutions of GTR which derive from the conservative view, for example the Friedmann models and the Schwarzschild solution, and we should remember that those established solutions provide us with a central core of information about gravitation.

Nevertheless, we should recognise that there is more to GTR than that which flows in the mainstream, and that even these 'central' features of GTR depend, to some extent, on less dominant aspects of the theory. For example, the cogency of the Friedmann models depends upon an examination of several matters which do not lie at the heart of GTR: for example the status of the cosmological principle, the empirical problems of redshift and of the microwave background, and the difficulties for GTR associated with initial and final singularities. But perhaps the principal reason for venturing outside the mainstream is that the conservative view of GTR does not enable us to describe a number of gravitational phenomena which we have good reason to believe occur in the *actual* world. Of course, we could wait for some 'unified' theory to displace GTR; but it would be more fruitful to do exactly what most physicists do and incorporate the best techniques available within the context of GTR, even if this means borrowing ideas from quantum theory. The very probable existence of singularities gives us every reason to do this, and, as we have observed, allows us to call into question any sharp distinction between 'classical' and 'quantum' approaches to gravitation.

Hence, if we take notice of the issues of §5.1, and decide to cater for opinion outside the mainstream, we have the chance to replace a relatively impoverished theoretical context with a wide-ranging set of ideas which facilitates the best available descriptions of gravitational phenomena and which recognises the important role of research at the 'periphery' of the theory.

I therefore propose that we should view GTR as a network of ideas which gives us a wide-ranging theoretical context, set within a theoretical framework. This context does not give us a 'static', definitive picture of GTR. For many of its components are relatively recent innovations. Hence, we should regard GTR as a *dynamic* or evolving theoretical context which embraces many different approaches, ideas and changes of direction.

Within the context of GTR, there seem to be six 'families' of ideas, namely those concerning

(i) the manifold and topology;
(ii) the metric;
(iii) the matter field;

(iv) the field equations;
(v) empirical data and evidence; and
(vi) the various principles of the theory.

And within each family of ideas we can identify:

(i) a mainstream or idea or set of ideas;
(ii) variations on that main theme or supporting ideas;
(iii) matters which are the subject of significant but reasonable dispute, or matters which are deemed to be of only minor concern.

For example, we might regard the mainstream view of the metric as demanding that it should be of Lorentz signature; variations on this theme may be that the metric may be Robertson–Walker, or that of the Kerr solution, and so on. Dispute may arise when we try to utilise a Lorentz metric in a contentious solution, e.g. the Gödel model. Again, we might regard the cosmological principle and the principle of general covariance/invariance as occupying mainstream positions. But we might decide either that the status of MP is the subject of significant debate, or that MP gives us a number of valuable theoretical variations, e.g. that due to Raine, which have a good deal of empirical respectability. † Obviously, the way we construct the complete picture of GTR's theoretical context will depend on the stage of development of the theory for the time at which the construction is attempted. For example, black holes, once unfashionable and highly contentious objects, are now quite respectable, if not firmly entrenched in the mainstream. Thus, we might present the current network of ideas in GTR's theoretical context as in table 5.1.

I do not intend that this sketch should be regarded as all-inclusive or as definitive. Indeed, much is absent from the columns for the variations and disputes. Rather, the table is meant to illustrate and give substance to the concept of a theoretical framework containing a dynamic theoretical context for GTR. Again, we should remember that any context constructed is liable to continual change as new evidence and theoretical thinking comes to light.

There are, perhaps, eight reasons why we should make the move from the narrow, conservative picture of GTR to this 'liberal', dynamic view.

(1) We can still give credit to the centre role of the mainstream position and its supporting ideas, i.e. to 'classical' GTR.

(2) We can also acknowledge the importance of the main

† We should note that it will not always be easy to distinguish the profitable variations of the main theoretical stream from the issues which are in dispute; MP, perhaps, gives us such a case.

Table 5.1 General Theory of Relativity: theoretical framework.

	Mainstream	Variations and supporting ideas	Disputes
Manifold and topology	Continuously differentiable, four-dimensional manifold	Uses of time-slices (achronal and Cauchy); singularity theorems	Closed causal loops
Metric	Lorentz signature	Robertson–Walker; Schwarzschild; Kerr; Taub–NUT; etc	Gödel solution; bimetric theories
Matter fields	The stress-energy tensor, $\boldsymbol{T}$	Additional scalar field (Brans–Dicke)	Negative pressure
Field equations	Einstein equation $\boldsymbol{G} = k\boldsymbol{T}$	Integral form: Einstein equation with Λ; use of quantum methods	Hoyle and Narlikar equation
Empirical issues and evidence	Equivalence; redshift; microwave background; homogeneity; isotropy	Presence of black holes	Presence of naked singularities
Principles	Principle of equivalence; principle of general covariance/invariance; and mass-energy conservation principles	Cosmological principle	Mach's principle

variations of the theory, which, like Raine's integral approach, have a high empirical standing.

(3) There are intimate links between empirical and theoretical aspects of the theory—empirical issues are *parts* of the theoretical context and influence directly the developments of the various approaches, e.g. the changing empirical pedigree of the cosmological constant has had a marked effect on the way we regard the field equations.

(4) Again, there are direct links between the principles of the theory and the theoretical context—for these principles have important effects on the development and on the contents of the theory, as we have seen in the cases of the principle of general covariance and the cosmological principle.

(5) There is no hard and fast distinction between 'classical' and 'quantum' approaches to GTR. Hence, we are able to avoid the notion that GTR is an isolated monolith.

(6) We can try to deal with all gravitational phenomena, and therefore we do not confine our attention to a restricted classical sub-set of phenomena. In particular, we can attempt to deal with singularities which, as we have observed, are consequences of 'classical' GTR.

(7) We can account for the continual changes in the theory, e.g. for the history and changing importance of the cosmological constant. Thus, we guard against the tendency of some to advance an artifical 'time-slice' of GTR as *the* correct version of the theory.

(8) We bring issues which are the subject of reasonable debate within the context of GTR. Such debates are part of the development of GTR and often have tremendous effects upon the mainstream of the theoretical context.

Like many others, Misner *et al* do not regard GTR-related research as providing a unified theoretical context. They imply that there are many different theories between which we may arbitrate on the basis of observation and experiments. Theories are reckoned to be worthy of serious attention so long as they are on an experimental par with 'classical' GTR (see Misner *et al* (1973: ch. 39)). But this view ignores the considerations of this and the last section. In particular it neglects the fact that 'classical' GTR has much in common with other approaches, as we have seen above, and it ignores the fact that GTR cannot stand on its own if it is to attempt the fullest possible description of the gravitational phenomena which arise from even just the 'classical' approach. The main elements shared by many of the various approaches to gravitation are:

(i) we are generally dealing with (Lorentz) metric approaches;

(ii) the Einstein field equations—though often modified—provide a central focus for work in the field;

(iii) every approach admitted into the theoretical context has good empirical and theoretical qualifications; and

(iv) the well established solutions also provide a focus for many approaches.

Hence, we can remedy the failings of the conservative view of GTR by adopting this more tolerant, 'liberal' view, which is wide-ranging, admits most GTR-related research into GTR's theoretical context and recognises the various relationships between the different components of the theory.

5.3 Kuhn and Feyerabend on science

Feyerabend notes two features of scientific practice which will help us to elucidate this tension between the conservative and liberal views of GTR. He sees (i) a tendency towards stubborn resistance—perhaps in the face of empirical evidence—in the defence of a theoretical viewpoint, and (ii) a tendency to develop ideas on a given theme quite freely even when these ideas challenge our theoretical prejudices. Feyerabend (1970) names the principles which underlie these respective tendencies as the principle of tenacity and the principle of proliferation. He is an advocate of proliferation at all stages of scientific development and he questions the way in which Kuhn's account of science and scientific theories seems to presuppose a rigid alternation between periods of tenacity and spasms of proliferation. His arguments also present a challenge to Lakatos' account, for this too seems to involve a commitment, if not so marked, to the alternation of tenacity and proliferation in scientific practice. Although Feyerabend addresses his remarks to Kuhn's discussion of paradigms and normal science, I shall take account of Kuhn's acknowledgment that the idea of a paradigm is not as well formulated as it might have been (see Kuhn (1970, 1974)).

A paradigm has been taken to refer to at least the central ideas, solutions and models of a theory which are likely to be encapsulated in one or more law-like relationships; hence the use of a paradigm enables the scientist to solve problems within a given domain of science. When the central ideas of a theory are well established, there is a period of normal science. When the theory runs into difficulties, a crisis may develop and new alternative ideas can then precipitate a scientific revolution leading to the adoption of a new paradigm (see Kuhn (1962)). It is this alternation between periods of normal science

and sudden changes of direction because of crisis and revolution which Feyerabend characterises as an alternation between tenacity and proliferation. Feyerabend's challenge is not diluted by Kuhn's introduction of the ideas of disciplinary matrices and exemplars. An exemplar is a standard example or solution—as so often displayed in text-books—which is part of the central core of information concerning a particular scientific theme. The Friedmann models of cosmology might be called exemplars. The disciplinary matrix is a grander idea. The matrix essentially fulfils the role assigned to the paradigm above; its advantage is that it lacks many other more specific roles orginally given to paradigms. The matrix accounts for the tenacity of scientists in holding fast to their exemplars, methods and practice. The breakdown of the matrix is responsible, in part, for the proliferation of new ideas which may result in a completely novel theoretical direction.

Lakatos' views also involve tenacity and proliferation. According to him, the scientist sticks firmly to a research programme and to the 'hard core' of the programme. In Newtonian dynamics the hard core includes the laws of motion and the law of gravitation, and, by the methodological decision of those engaged in the research programme, these laws are 'irrefutable', so long as that programme is pursued. Only a new programme can overthrow or falsify the hard core of the old programme. This can happen when the old programme becomes less and less fruitful. Then 'bold conjectures' may lead to the adoption of a new programme. Thus, the research programmes give us periods of tenacity, and the conjectures give us proliferation (see Lakatos (1970)).

This picture of science and its development arises out of Lakatos' attempt to give Karl Popper's 'falsificationism' a more structured and historically accurate format, and to exorcise some of its more problematic features. Popper's influential ideas show great dissatisfaction with the use of induction as the methodology of science (see especially Popper (1959, 1963)). He argued that we can never prove a theory to be true, no matter how many confirming instances we supply. For we will always rely on an induction over possibly falsifying instances. Hence Popper asks scientists to adopt a more rigorous attitude: they should aim to find the weak points of theories and should look for the *falsifying* instance which would overthrow the theory. Although no confirming instance can provide us with proof, a falsifying instance is enough to disprove the theory in question. Although Popper's ideas are superficially attractive, there are many difficulties arising from his approach.

(1) Scientists rarely overthrow a theory simply because of falsifying instances. Hence Popper's approach seems unsatisfactory, if only from a historical perspective.

(2) The claim that a falsifying instance has disproved a theory seems to involve inductive thinking. How can we be sure, for example, that a falsification today will remain so tomorrow, and why do we have such confidence in the apparatus used in the experimental test—perhaps because it has performed well in the past? Indeed, Nelson Goodman argues that all our language is riddled with induction: the claim that an emerald is green, for instance, involves a denial that it will not suddenly change colour at some future time—and why should we believe the claim except given the inductive reasoning that we do not observe such objects changing colour in our (relatively) stable environment?†

(3) Popper's advice to scientists is to formulate bold conjectures ‡ and then to seek a refutation with rigorous testing. But this depends on a 'pure' oriented view of science. How are theories to develop sufficiently to allow technological advances? If we follow his advice strictly, then applications would arise by default simply because a theory is not yet falsified.

Lakatos provides at least a preliminary response to these worries. By arguing that scientists need a new theory before they are willing to regard the old theory as falsified, he believes that the first difficulty is overcome. Hence, falsificationism, given his historical perspective, can account for the unwillingness of scientists to disown a theory in the face of apparently damning evidence. Because Lakatos maintains that scientists should adopt a range of empirical data and beliefs by methodological decision this allows us to test a theory and not rely on inductive thinking. When we say that our apparatus is trustworthy, we do so because of a methodological decision, not because of any inductive faith in its performance. Finally, because Lakatos sees tenacity as a natural inclination of the scientific community, he is able to defend himself against the accusation that falsificationist methods leave no room for the development of applied science. Despite Lakatos' embellishments, falsificationism nevertheless seems to rely on an assumption of the relative stability of the physical world. For how can we decide that a certain collection of facts is to be treated as if the facts were correct, unless we regard the world as being a guarantor of enough stability for us to treat at least *some* facts as correct? Of course, it seems irrational for anyone to doubt that the universe is relative stable and simple in its behaviour and structures. But it is exactly this lack of doubt which underwrites the inductive approach to science. This theme is pursued by many critics of falsificationism, (see, in particular, Mackie (1980) and Hesse (1974, 1980)). Hence we

† See Goodman (1983) for a full discussion of his 'Grue' paradox.

‡ Note that Popper allows inductive thinking when we are 'dreaming up' new theories, but not when we are testing them.

should be content to note that we should beware of any naive acceptance of the falsificationist programme, even in the sophisticated form presented by Lakatos.

Despite the many differences between Lakatos and Kuhn, the ideas of the hard core and the disciplinary matrix share two features: first, they both characterise periods of tenacity in research when scientists are extremely reluctant to depart from established paths and, secondly, they are both distinguished by the role they accord to the 'fundamental' laws of a theory. Lakatos regards such laws as being 'immune to falsification', at least for the duration of the research programme to which the laws belong. Kuhn maintains that the laws or 'symbolic generalisations' at the heart of the matrix are the defining characteristics of the discipline concerned—hence the adoption of a new set of laws must signify a change of matrix. Therefore, for Kuhn too, basic laws seem to be immune to revision whilst the matrix holds sway.

Feyerabend says that Kuhn's picture of normal science is not supported by the facts of history. He maintains that we can identify periods of allegedly 'normal' science—according to Kuhn's criteria—in which several distinct and mutually incompatible approaches to a scientific issue actively combine to produce new directions in science. He gives the field approach of Faraday and Maxwell and the mechanical approach evident in Newtonian dynamics as an example of this. He claims that

> The troubles leading to the special theory of relativity could not have arisen without the tension that existed between Maxwell's theory on the one side and Newton's mechanics on the other. (1970: 208)

Thus, Feyerabend sees the proliferation of ideas as an essential ingredient of a scientific crisis. But he insists that proliferation is also an important ingredient of science as a whole, going beyond the bounds which Kuhn allows for 'proliferators'—a dissident minority—during the rule of normal science:

> Proliferation does not start with a revolution; it precedes it . . .. Science as we know it is not a temporal succession of normal periods and of proliferation; it is their juxtaposition. (1970: 212)

Feyerabend proposes that tenacity and proliferation are *not* temporally separated, but co-present throughout the development of a theory. We should regard tenacity in science as a commitment to follow, develop and defend one's own ideas; and the principle of proliferation entails that 'everyone may follow his inclinations' with the result that science is the immediate beneficiary (see Feyerabend (1970: 210)).

Feyerabend says that he sees this synthesis of tenacity and proliferation in Popper's view that science advances by the critical comparison of alternative views. He also notes Lakatos' belief that work on theoretical ideas outside the main research programme in hand is likely to be required for the purposes of such comparisons. Lakatos says that he welcomes competition between alternative research programmes, and that 'the sooner competition starts, the better for progress' (1970: 155). Accordingly, he welcomes Feyerabend's view of 'theoretical pluralism' as opposed to Kuhn's 'theoretical monism'. But we should remember that Lakatos' position on this matter is not too far from that taken by Kuhn. Although Kuhn emphasises the role of normal science and plays down the idea of competition between alternative views whilst normal science is in session, his fundamental *commitment* is to a disciplinary matrix which, so long as his commitment lasts, is in certain respects immune to revisions. Lakatos may disagree with Kuhn about the role of normal science and its overall dominance of the scientific scene, and he may push the Popperian view that competition is healthy, but he too regards the basic ideas of a research programme as sacrosanct until degeneration sets in and a new programme begins.

Feyerbend does not take exception to Lakatos' reading of history, given that Lakatos allows other programmes to detract from the dominance of any single, well established programme. But Lakatos also believes that research programmes do require tenacity: this is shown by his contention that the basic tenets of a programme, the elements of the hard core, are immune to revision. Hence, the workers on that programme are discouraged from actively engaging in the proliferation of ideas within the context of their own programme. Therefore I believe that we can apply Feyerabend's analysis to Lakatos as well as to Kuhn: both emphasise the need for tenacity in holding on to the fundamental ideas of research programmes and disciplinary matrices respectively and both maintain that, for the scientists involved in a particular programme or matrix, proliferation only comes into its own at times of degeneration or crisis. Consequently, Feyerabend ought to be unhappy with Lakatos' view also. For Kuhn *and* Lakatos deny, in a fundamental way, the *complete* synthesis of tenacity and proliferation which Feyerabend advocates. Feyerabend's belief that we can all follow our inclinations in science implies that even the workers on a given programme or in a particular discipline may do as they will. But if 'anything goes', there is no reason to hold on even to the basic ideas of a theory. Therefore the hard core and the elements at the heart of the matrix are obviously endangered species: they would lose the protected status which Kuhn and Lakatos give to them, that is if Feyerabend has his way.

I believe that my examination of the content and structure of GTR gives some support, and adds a good deal of substance, to Feyerabend's analysis. But I do not believe that it underwrites his prescription that in science 'anything goes'. The picture given of GTR is indeed one in which tenacity and proliferation are (reasonably) happy bedfellows. On the other hand, the central role of the mainstream, and in particular the tendency for those involved with GTR to focus their attention on the constituent elements of the mainstream is acknowledged. Yet much theory-related research and many ideas, apparently at odds with the contents of the mainstream, are admitted into GTR's theoretical context. The range of this theoretical context is certainly wide enough to cater for the synthesis of tenacity and proliferation: from the well established models, such as Schwarzschild's solution, to the imaginative marriage of 'classical' and 'quantum' methods which is put to good use in the work on singularities.

Of course, the proliferation of some ideas might be regarded as being well within the bounds of what Kuhn calls normal science, with little threat to the basis tenets of the theory. This is obviously correct. The development of new 'classical' solutions and the work on the properties of Cauchy hypersurfaces, for example, seem to fall into this category. Such proliferation is innocuous as far as workers committed to Kuhn's disciplinary matrices or Lakatos' research programmes are concerned. But, the proliferation of ideas observed within the context of GTR goes well beyond the safety of these alleged boundaries. Two examples stand out.

(1) We allow into GTR's theoretical context what Kuhn would call 'revolutionary' ideas—namely those of quantum theory. That is, we admit ideas which are revolutionary from the perspective of those who follow classical approaches to gravitation.

(2) We are prepared to accommodate revisions to the fundamental tenets of the theoretical context *during* its lifetime. Indeed, even the basic laws of the theory are not safe from modifications—for example the inclusion of the comological constant, the addition of a scalar field, or the move to an integral form for the field equations.

In the first case, we disrupt the continuity essential for Kuhn's alleged normal science; and in the second case, we allow rigorous challenges to and even revision of the basic ideas of a theory which neither Kuhn nor Lakatos are willing to entertain.

But the fact that proliferation is allowed during the lifetime and within the context of GTR does not mean that 'anything goes'. Feyerabend seems to view proliferation in a fairly unstructured way, and places no obstacles before the scientist who seeks to develop his ideas. He certainly does not argue that proliferation occurs in a

relatively controlled manner, and that the initiation and development of ideas is constrained by the theoretical context and its place in science as a whole. But this is how the development of GTR forces us to view the proliferation of ideas. For example, singularities of collapse only became 'respectable' when they acquired a theoretical pedigree with empirical ramifications *within* GTR's context. But naked singularities have yet to acquire such theoretical force, and so remain, at best, on the borders of the context. Closed causal curves are deemed to be highly implausible entities, and are perhaps only of instrumental value. Perhaps, too, the idea of non-metric space–times may be regarded as being outside GTR's context. I do not believe that any hard-and-fast rules are in operation here; but it is evident that the liberal view of GTR is not as anarchic as Feyerabend might wish. Certainly, it is not a matter of the *free* choice of the scientific community, for they are constrained by the conceptual demands of the theoretical framework at any stage of the development of GTR.†

Nevertheless, like Feyerabend, I do not believe that normal science, as characterised by Kuhn, exists. Nor do I think that the idea of a hard core is particularly valuable, especially since Lakatos fails to tell us how we can tell *when* we know we have a hard core. In an illuminating footnote, Lakatos admits that

> The actual hard core of a programme does not actually emerge fully armed like Athene from the head of Zeus. It develops slowly, by a long, preliminary process of trial and error. (1970: 133)

But how and when in the life of a programme does this slow development take place? And how can we tell that we have reached the end of the development? Lakatos might say that the development of GTR was not complete until after:

(i) the publication of a 'definitive' version of the programme (e.g. for GTR after the publication of Einstein's 'The foundation of the General Theory of Relativity' in 1916) or,

(ii) the first major corroboration of the programme's novel predictions (e.g. perhaps after the time of Eddington's confirmation of GTR in 1919).

It is certainly difficult to gauge the time at which a theoretical context has settled down enough to allow the identification of a hard core. If we choose 1916 for GTR, then part of that hard core must be the original Einstein field equations. But if we choose 1919, the hard core will include the modified equations which involve the use of the cosmological constant and a commitment to the idea of a sourceless

† I shall return to the problem of how scientists control the development of their theories in §§5.4 and 5.5.

field. If we wait a little longer, perhaps until 1929, when the idea of an expanding universe became empirically respectable, then we might decide that the hard core does not include the cosmological constant. Yet, as is evident from the recent work on the early universe presented in Gibbons, Hawking and Siklos (1985), the constant is far from a dead issue. So, even if we take a narrow, conservative view of GTR, we could not say with certainty that one particular version of the field equations is a part of the hard core.

Perhaps the only option available to us, on a conservative view, may be to claim that Will's 'superstrong' equivalence principle (SEP) is such an essential feature of GTR that SEP is the hard core. As we found in §2.1 (see p. 61*f*), the adoption of SEP seems to isolate metric theories and possibly picks out only 'classical' GTR. However, as we have observed (in §2.2), strong equivalence principles are not *strictly* true, and must be regarded as *heuristic* devices. Hence, although SEP and similar principles may point us towards GTR, they cannot properly be said to be part of the theory. Moreover, if it is to have any substance, the idea of a hard core ought to involve the demand for a unique set of laws. But SEP is compatible with not only the field equations with and without the cosmological constant, but also Raine's integral approach, as well as other possible variants, as Will himself acknowledges.

The possibility of picking out a 'permanent' hard core seems to be negligible: any such attempt would seem to be a *post hoc* characterisation of one particular line of thought. On a liberal, dynamic view of GTR, the case for a hard core becomes even weaker. For any *post hoc* characterisation of a theory would neglect the dynamic evolution of the theoretical context and its laws. Hence, not only is is difficult to argue for a hard core for GTR, to do so would be quite misleading, for the idea of a hard core relies on a static view which is challenged strongly by the historical development of GTR.

5.4 Relativity—a dynamic, evolving theory

We have contrasted the static, conservative view of relativity with a dynamic and more liberal perspective. This liberal view presents a challenge to those who, like Kuhn, would like us to see a period of calm between the stormy revolutions which mark wholesale theory changes. I can understand why philosophers and historians of science are attracted by Kuhnian descriptions of science. The characterisation of theories and their development becomes a reasonably straightforward matter, with the central laws of a theory playing a

fundamental role. With references to 'standard problems', Kuhn's approach is very much in line with a text-book view of science. It is almost as if the essential ideas of a theory are regarded as lifeless specimens. The real life in the theory consists of solving puzzles and problems, that is in how scientists *use* the essential ideas. The theory is treated as if it were complete. Any problem which we cannot resolve is set on one side as anomalous. To challenge the essential ideas of a theory would be to demand a revolution and a new theory.

This tidy view would be fine, if it were not for the fact that scientific research is often problem-orientated. In order to resolve an empirical or theoretical problem, theories are adapted and contexts are enlarged to provide the necessary resources. Kuhn sees a dramatic divorce between the classical physics of Newton and quantum theory. Yet, as we have seen, in order to investigate the properties of singularities, classical *and* quantum views are used together. Despite the fact that, on a Kuhnian view, classical and quantum physics are radically different languages, a composite of these approaches can and does resolve problems. Again, the problem of Mach's principle has evoked a variety of responses which have collectively enriched the theoretical context of GTR. Relativists have not simply set this problem on one side as anomalous because Einstein's original equations do not validate the principle; rather, they struggle to adapt the theory or to produce new variations which might resolve the problem.

Philosophers often refer to theories such as Newtonian mechanics as if they were 'dead': they pick out what they regard as the key elements of the theory and pronounce that, given Newtonian mechanics, such and such must follow. Whilst they might have some doubts about their philosophical reasoning, they have few worries about the nature of Newtonian mechanics: after all, don't we all know what the essential features of the theory are? Kuhnian perspectives help to promote such views.

It is interesting to contrast the attitude of philosophers to relativity theory with their approach to quantum mechanics or particle physics. There is little uncertainty displayed about the nature of relativity theory in the extensive literature generated by philosophers of science. Problems invariably arise in the context of what are perceived to be the philosophical implications of the theory. † Indeed, many such problems have been discussed in the preceding chapters. But philosophers approach quantum mechanics much more nervously because, here, it is harder to identify a single set of equations which can be neatly pigeon-holed as 'the laws of quantum mechanics'. Hence, philosophers of quantum mechanics are much more likely to

† Friedman (1983) is a welcome exception to the general rule.

be concerned about foundational issues (see, for example, Jammer (1974)). They are more inclined to ask what quantum mechanics actually is, rather than leap-frog to the implications of the theory. They may be prompted by the fact that there is not *one* quantum mechanics, but several distinct views. And when we consider the relative chaos of particle physics, it is hardly surprising that little is said by philosophers. Perhaps there is just too much confusion and certainly no tidy parcel of 'laws' (see Pais (1986)). Particle physics, especially, appears to be very much 'alive', with much to be settled.

But relativity is just as 'alive' as particle physics. Any tendency to regard a theory as 'dead' derives from a narrow attitude of mind rather than from historical and scientific facts. Many might regard Newtonian mechanics as 'dead'. But, as Truesdell (1984) points out, this may be due to the misapprehension that classical mechanics is no more than analytical mechanics, and concerned only with mass points, rigid bodies and perfect fluids. Mechanics, however, has moved on: bodies may have microstructures, motions need not be smooth, forces may be complex, and additional equations add to the generality of the subject. Again, electromagnetism has had to extend itself beyond its classical foundations with concern only for macroscopic phenomena. Maxwell's equations assume that matter has no discrete microstructure. Hence, as Robinson (1973) remarks, if classical electromagnetism is to be coherent, questions about the nature of elementary charged particles must be evaded, But many electromagnetic phenomena cannot be properly understood without an atomistic perspective, for example the phenomenon of dispersion (see Bleaney and Bleaney (1976)). Therefore the same kind of marriage already seen between classical and quantum views in relativity is demanded in electromagnetic theory. The theory cannot stand still. Indeed, it may always be a mistake to regard a theory as 'dead': evolution and change seem to be demanded if the physical world is to be properly understood. A dynamic theoretical context permits such changes to take place without the immediate need for any dramatic revolution.

One reason for the eagerness to characterise theories as relatively static organisms may be the pre-eminence given to the role of scientific laws. We have already noted that laws are often regarded as the incorrigible focal points of a theory. But recent arguments, by Cartwright (1983) and also Hacking (1983), challenge this special status given to theoretical laws. They maintain that the important laws of physics are the 'phenomenological' laws which help us to understand specific experimental problems. Such a law is often derived from a pastiche of theories. For Nancy Cartwright theoretical laws are not strictly true in any *real* circumstance. So why should we accord these laws such respect? She asks us to believe in theoretical

entities rather than theoretical laws: the former can be utilised and manipulated to deliver the experimental 'goods', the latter are merely instrumental aids, certainly useful in their abstract domains, but with no straightforward *ontological* significance. † If anything, the moral to be drawn from this attack on theoretical laws is that we should avoid the belief that such laws are unproblematic. And this may make us think twice about giving any particular law a pre-eminent status within a theoretical context. Certainly, the context of relativity theory demonstrates that we cannot do so in any simplistic way.

A further lesson can be learnt from Cartwright and Hacking: we should beware the tendency to regard any theory as an isolated monolith. We might recall that Lucas views GTR as a monolithic structure.‡ But realistic problems and phenomena may require a broad approach utilising, for example, both classical and quantum methods, as required in the case of singularities. The enlarged theoretical context of GTR is not simply an artificial construction, it provides us with the means to tackle realistic gravitational problems. We can do so in a variety of ways using a wide range of approaches, whether from the mainstream or the periphery of the context. Without this enlarged, dynamic context, we are forced to the conclusion that GTR *per se* is of limited value, just as is the case with classical electromagnetism and analytical mechanics. Such restricted theories are the true artificial constructions: the stuff of elementary text-books. Their ranges of application may be interesting and instructive, but if they are to deal with the fullest possible range of realistic phenomena such theories need the 'life' brought by an evolving, dynamic theoretical context.

In the last section, I maintained that a liberal view of GTR does not mean that there should be *no* constraints on the theoretical context. I suggested that the constraints are largely conceptual and are based on the adherence of the community of researchers to the conceptual demands of the theoretical context. Can this suggestion be given a little more substance? For it may not be clear why conceptual constraints on the proliferation of ideas (within the theoretical context) produce a reasonable degree of order and coherence in our picture of gravitation *qua* GTR. We know that collective decisions seem to have been taken to admit such and such an idea, to regard other ideas as borderline cases and to dismiss yet others. Can we justify

† A possible response to this view is to claim that although the present laws of physics are strictly false, research will lead us to the true set of laws. But such a response poses many difficulties which require a more detailed discussion than can be given here (see Newton-Smith (1981) and Laudan (1986)).

‡ See Lucas (1973) and p 157 above.

these decisions? My initial answer to this problem has been that relativists will tend to exclude ideas, models, laws etc, the inclusion of which would result in radical conflicts with the context or which have no empirical or theoretical validity from the perspective of GTR's theoretical context. Of course, gradual changes in the context may eventually produce conditions ripe for the admission of ideas which were previously taboo. But we need to advance beyond these preliminary responses.

In *The Structure of Scientific Inference* (1974), Mary Hesse presents an account of the network model of scientific theories, firmly based on the legacy of Duhem and Quine. An examination of her account shows that the dynamic theoretical framework of GTR is consistent with the network model, and that one of the reasons for the orderly progress of research in GTR may be found in Hesse's 'coherence conditions' (see, in particular, pp 9–37 of Hesse's book). The second and fundamental reason for orderly progress in research is discussed in §5.5.

The presupposition of Hesse's account is that we possess a basic conceptual ability to pick out and classify similarities and differences amongst the data we meet in experience. This provides us with a network of classifications—a network of predicates and their interrelationships which is firmly embedded in the domain of observations. But how do we order the vast number of classifications presented to us in our experience? Judgments of similarity and of difference between elements of our experience would only give us a very loose grip on the physical world. We can make progress, however, by a process of grouping and re-arranging the initial classifications of data with the consequence that the basic interrelationships between the elements of the network are refined and shaped into laws. Hesse suggests that a reprocessing of the initial classification of data takes place according to coherence conditions. The decision to impose such conditions is a conventional matter, and results from the desire of the scientific community for relative order from the chaos of experience. This decision involves a commitment to hold fast to certain elements of the network, to certain classifications and laws.

We can now arrange a marriage between Hesse's network model and the dynamic theoretical framework of the GTR. The context of GTR is indeed one in which the various elements—facts, laws, models, principles—form a network of interacting ideas and relationships within the framework outlined above. And the process of reclassification of the original data is certainly evident, e.g. in the decision to adjust the Einstein field equations to accommodate Einstein's belief that the equations of GTR should not preclude a static model. It is also clear that the ideas in the mainstream of the framework, together with some of the major, well established variations, represent those elements to which physicists feel particularly attached; and these are

their coherence conditions for GTR. Hence research in accordance with those conditions may produce the order and coherence desirable in scientific theories.

We have acknowledged that no element within the theoretical context is immune to revision. Hesse certainly agrees with this Quinean sentiment when she says that

> the network inter-relations between more directly observable predicates and their laws are in principle just as subject to modifications from the rest of the network as are those that are relatively theoretical. (1974: 24)

But if *all* the elements of the context are subject to revision, then, contrary to Hesse's wishes, the coherence conditions of the network may therefore be revised. Such revisions may well take place gradually, especially when they concern the more dominant coherence conditions; but there is no doubt that, in GTR, these revisions do occur. A good example is the increasing dominance of the Robertson–Walker solutions, particularly the cosmological Friedmann models. This goes hand-in-hand with the gain in respectability for the empirical evidence which suggests that the universe is expanding from an initial singularity. Again, several other aspects of GTR, which play a central role in ordering our views on gravitation, have been subject to revision over the years: for example the Einstein equations themselves and our basic views of the causal structure of GTR. Accordingly, we should emphasise that the dynamic nature of GTR's theoretical context requires a thoroughly dynamic view of the network model, with changes, not just in the secondary classifications and regularities, but also in the primary coherence conditions.

We should also recognise the role of principles within the framework of GTR. For these, too, have an important influence on the nature and development of the theoretical context, and they are firmly tied to many of the other elements within the framework. For example, the cosmological principle has had some influence in the rise in status of the Friedmann models, and is attached to various empirical facts concerning homogeneity and isotropy, to the Friedmann models themselves, to various modifications of the field equations, e.g. that due to Raine, and, via the Friedmann models, to Mach's principle. Similarly Mach's principle has strong ties with relationist solutions, particularly the Friedmann models, and the approaches of Brans and Dicke and of Raine.

Whilst the network model enables us to adopt a model of the theory as a coherence condition, it does not allow us to do this in isolation. This fact has direct consequences for the relationist who wishes to focus our attention on solutions with the Robertson–Walker metric

and who attempts to restrict the theoretical context to those elements which are in accord with such solutions. This, of course, may be seen as the relationist's prescription for Mach's principle. But such a view would ignore the dominant role played in the network by such solutions as that due to Schwarzschild. Although the liberal view allows the admission of MP into GTR's context—because of the compatability of this principle with various other elements of the context—it would be cavalier to assert that physicists are so confident in the empirical evidence which supports the predilections of the relationist that they regard Mach's principle as not just part of the mainstream but a dominant feature. Of course, the context may change to tip the scales in favour of Mach's principle. But until then we cannot say that GTR vindicates Mach's principle; but we can nevertheless maintain that GTR involves this principle. Hence, there is room for both relationism and absolutism in GTR.

5.5 Simplicity and scientific theories

The network model outlined in the previous section has much to recommend it. The flexibility it offers to our description of the scientific enterprise certainly accords with our account of GTR, and it would seem eminently suitable for contexts such as particle physics. But there remains a fundamental problem. The network model, in the form I have adopted, allows the revision of any element in a theoretical context. We must therefore characterise exactly what prevents scientists from changing a context so much that we effectively reach some anarchic conclusion. Feyerabend's dictum, anything goes, might well rule. And this clearly doesn't happen in GTR or any other entrenched scientific context. Hence, we need to uncover the conceptual controls which govern the orderly progress of research within a context and which ensure the overall coherence of the theory. The key to this problem is found in the search for simplicity. For the various demands of simplicity ensure not only internal coherence but also consistency with more general, external scientific constraints.

The search for simplicity has certainly affected the development and structure of GTR. This is especially evident in the following.

(1) The principles of equivalence, particularly when equivalence is used as a heuristic principle of minimum coupling.

(2) The principles of covariance and invariance, which make strong demands on the form and context of GTR.

(3) Mach's principle, when interpreted as a methodological injunction to rule out absolute space–times.

(4) The cosmological principle, which asks us to adopt the simple physical structures of homogeneity and isotropy for the actual universe.

(5) The support given to the metric of STR as a basic form for the gravitational field.

(6) The adverse reactions against variations of GTR which postulate additional structures such as scalar fields.

(7) The demands for topological constraints which rule out such peculiarities as space–times in which time direction is not well defined locally, and so prevent the theoretical context from becoming unduly pathological.

(8) The demands for constraints on the matter fields of GTR, which asks us to avoid such problematic possibilities as negative local kinetic energy, which, as well as being empirically questionable, also add unwanted complexities to the context of GTR.

(9) Finally, the idea of cosmic censorship which attempts to simplify the difficult problem of singularities and to maintain as much predictability as possible in GTR space–times.

I believe that the coherence conditions of a theoretical context must include such simplifying principles and constraints if maximal coherence is to be achieved. For the various demands of simplicity limit possible extravagances in form and content and help us to avoid any complexities which might prevent the context from yielding useful results. Without such constraints the theory could indeed be 'enriched' to the point of uselessness: to be too general to have anything of specific value to say about the world. Hence, the search for simplicity provides the essential conceptual controls on the fruitful and coherent developments of GTR and, I would suggest, of all scientific theories.

The search for simplicity may lead us into a possible trap. When simplicity is used as a basis for methodological constraints on the structures and development of a theory, there will be positive gains in the coherence of that theory. But if we go on to claim that the physical universe *demands* a description and explanation dependent upon criteria of simplicity, then we are making strong ontological claims which require justification. Considerations of conceptual simplicity, symmetry and elegance in formal presentation may be powerful instrumental aids, but we would have to adopt a realist philosophy if we wish to say that the simplicity is indeed a sign of truth. What I have said on the use of simplicity as a control on the structures and development of GTR does not force us to adopt either a realist or antirealist perspective. Simplicity is primarily a methodological injunction, and only after careful philosophical consideration should we consider making it a metaphysical fiat.

Appendix

Ellis, in his 'Is the Universe expanding?' (1978) maintains that we can drop the cosmological principle, build an inhomogeneous *static* model and still account for all the observations which cosmologists cite in favour of the particular kind of simplicity involved in Friedmann models. Ellis says

> The usual deduction that the universe is expanding is made on the basis of observations of the systematic red shifts of galaxies which are interpreted as cosmological Doppler shifts, and is supported by the interpretation of the microwave background radiation as the relic radiation from a hot big bang (1978: 87).

Ellis then maintains that

> . . . these observations can also be explained by a static spherically symmetric universe model with two centers, and our galaxy near one of the centers. In this case the systematic red shifts of the galaxies are interpreted as cosmological gravitational red shifts, while the microwave background radiation originates from a hot gas surrounding a singularity situated at the second center of the universe. (1978: 87).

Ellis maintains that there is a close analogy between his model and the Friedmann models

> For virtually every problem encountered in the one universe there is a corresponding problem in the other. (1978: 91).

Ellis admits that we could discover an observational difference between the two types of model, † and he agrees that all the details of his cosmological picture are not altogether clear. But he asks why we

† For example, the observation of the time variation of some cosmological quantity like the Hubble constant (1978: 93).

have not investigated the possibilities of such static models before now. He says

> Models of the sort described here have not been considered previously because of the assumption—made at the very beginning in setting up the standard models—of a principle of uniformity. . .. This is assumed for a priori reasons and not tested by observations. However, it is precisely this principle which we wish to call into questions. (1978: 92).

It is clear that once we assume that cosmological principle, if our models are to be empirically adequate, we must make them expanding models.

> This assumption is made because it is believed to be unreasonable that we should be near the center of the universe. (1978: 92).

So our 'observation' that the universe is expanding is connected with our Copernican belief that 'here is just as good as anywhere'. Ellis replies to this:

> [It] is certainly unreasonable if the implication is taken to be that the universe has been centered on our presence; however, there is no need for this implication. Rather one should ask: Given a universe model of the type proposed, where would one be likely to find life like that we known on earth? (1978: 92).

Ellis' model has two centres, that with the singularity is very hot and that without is reasonably cool. Life will occur in this cool situation if anywhere. It may be argued that life only occurs at certain, favourable times in the history of a Friedmann universe. We would not expect life to occur at early (hot) or late (cold) times. Ellis' model gives a spatial analogy to this: life occurs in specific places in space, rather than at specific times. The claims which Ellis make for his model are modest; he says

> It is not claimed that the universe is actually like this model. What is claimed is that there are no overwhelming arguments that immediately show that such a model could not reproduce all the current observations. (1978: 93).

But the implicit message is clear. Ellis warns us lest our metaphysical prejudices overwhelm our critical good sense.

References and Bibliography

d'Abro A 1927 *The Evolution of Scientific Thought* (New York: Dover)
Adler R, Bazin M and Schiffer M 1975 *Introduction to General Relativity* 2nd edn (New York: McGraw-Hill)
Alexander H G 1956 *The Leibniz–Clarke Correspondence* (Manchester: Manchester University Press)
Anderson J L 1964 Relativity principles and the role of co-ordinates in physics in *Gravitation and Relativity* ed H Y Chiu and W F Hoffman (New York: Benjamin)
——1967 *Principles of General Relativity* (New York: Academic)
——1971 Covariance, invariance and equivalence: a viewpoint *Gen. Relativ. Grav.* **2** 161–72
Anscombe G E M 1975 Causality and determination in *Causation and Conditionals* ed E Sosa (Oxford: Oxford University Press)
Armstrong D M 1983 *What is a Law of Nature?* (Cambridge: Cambridge University Press)
Barbour J B 1982 Relational concepts of space and time *Br. J. Phil. Sci.* **33** 251–74
Barrow J D and Tipler F J 1986 *The Anthropic Principle* (Oxford: Oxford University Press)
Bath G T (ed) 1980 *The State of the Universe* (Oxford: Oxford University Press)
Bernstein J 1973 *Einstein* (London: Fontana)
Birch P 1982 Is the Universe rotating? *Nature* **298** 451
Blackmore J 1972 *Ernst Mach* (Berkeley: University of California Press)
——1985 A historical note on Ernst Mach *Br. J. Phil. Sci.* **36** 299–305
Bleaney B I and Bleaney B 1976 *Electricity and Magnetism* (Oxford: Oxford University Press)
Bohm D 1964 *Causality and Chance in Modern Physics* (London: Routledge and Kegan Paul)
Bondi H 1967 *Assumption and Myth in Physical Theory* (Cambridge: Cambridge University Press)
Bosclough J 1984 *Beyond the Black Hole* (London: Fontana)
Bradley J 1971 *Mach's Philosophy of Science* (London: Athlone)

Brans C and Dicke R H 1961 Mach's principle and a relativistic theory of gravitation *Phys. Rev.* **124** 925–35
Bridgman P W 1936 *The Nature of Physical Theories* (New York: Dover)
Brown H R and Weingard R 1988 *Einstein* (Oxford: Oxford University Press)
Büchner L 1884 *Force and Matter* (London: Asher & Co.)
Bunge M 1963 *The Myth of Simplicity* (Englewood Cliffs, NJ: Prentice-Hall)
——1966 *Foundations of Physics* (Berlin: Springer)
——1973 *Philosophy of Physics* (Dordrecht: Reidel)
Capek M 1961 *The Philosophical Impact of Contemporary Physics* (New York: Van Nostrand)
——(ed) 1976 *The Concepts of Space and Time* (Dordrecht: Reidel)
Carnap R 1936 Testability and Meaning *Phil. Sci.* **3** 419–50
Cartwright N 1982 *How the Laws of Physics Lie.* (Oxford: Oxford University Press)
Chalmers A F 1982 *What is This Thing Called Science?* (Milton Keynes: Open University Press)
Chiu H Y and Hoffman W F (ed) 1964 *Gravitation and Relativity* (New York: Benjamin)
Clarke C J S 1977 Time in General Relativity in *Foundations of Spacetime Theories* ed J Earman, C Glymour and J Stachel (Minneapolis: University of Minnesota Press)
——1976 Space–time singularities *Commun. Math. Phys.* **49** 17–23
——1979 *Elementary General Relativity* (London: Edward Arnold)
——1982 Singular space–time *Commun. Math. Phys.* **84** 329–31
Cohen R S and Seeger R J (ed) 1970 *Ernst Mach* (Dordrecht: Reidel)
Collins C B and Hawking S W 1973 Why is the Universe isotropic? *Astrophys. J.* **180** 317–34
Colodny R (ed) 1986 *Pittsburgh Studies in Philosophy of Science* (Pittsburgh: Pittsburgh University Press)
Davidson D 1980 *Essays on Actions and Events* (Oxford: Oxford University Press)
Davies P C W 1974 *The Physics of Time Asymmetry* (Guilford: Surrey University Press)
——1980 *Other Worlds* (London: Dent)
——1986 *The Forces of Nature* (Cambridge: Cambridge University Press)
DeWitt B S 1983 Quantum gravity *Scientific American* **249** 74–85
Dicke R H 1964 The many faces of Mach in *Gravitation and Relativity* ed H Y Chiu and W F Hoffman (New York: Benjamin)
Dolgov A D 1985 An attempt to get rid of the cosmological constant in *The Very Early Universe* ed G W Gibbons, S W Hawking and S T C Siklos (Cambridge: Cambridge University Press)
Dorling J 1978 Absolute space *Br. J. Phil. Sci.* **29** 311–23
Duhem P 1954 *The Aim and Structure of Physical Theory* (Princeton: Princeton University Press)
Dummett M 1978 *Truth and Other Enigmas* (London: Duckworth)
Eardley D M and Smarr L 1978 Time functions in numerical relativity; marginally bound dust collapse *Phys. Rev.* D **19** 2239
Earman J 1970 Who's afraid of absolute space? *Aust. J. Phys.* **48** 287

——1970 Space-time, or how to solve philosophical problems *J. Phil.* **67** 259–77
——1971 Laplacian determinism, or is this any way to run a universe? *J. Phil.* **68** 729–44
——1972 Causal propagation outside the null cone *Aust. J. Phys.* **50** 222–37
——1976 Causation: a matter of life and death *J. Phil.* **73** 5–25
Earman J and Friedman M 1973 The meaning and status of Newton's law of inertia *Phil. Sci.* **40** 329–59
Earman J and Glymour C 1978 Lost in the tensors *Stud. Hist. Phil. Sci.* **9** 251–78
Earman J, Glymour C and Stachel J 1977 *Foundations of Spacetime Theories* Minnesota Studies vol. VIII (Minneapolis: University of Minnesota Press)
Eddington A 1920 *Space, Time and Gravitation* (Cambridge: Cambridge University Press)
——1934 *New Pathways in Science* (Cambridge: Cambridge University Press)
——1979 Gravitation and the principle of relativity *Nature* **278** 213
Einstein A 1905 On the electrodynamics of moving bodies (in Einstein A, Lorentz H A, Weyl H and Minkowski H 1923 *The Principle of Relativity* (New York: Dover))
——1909 Letters to Mach, 9 August and 17 August
——1911 Influence of gravitation on the propagation of light (in Einstein A , Lorentz H A, Weyl H and Minkowski H 1923 *The Principle of Relativity* (New York: Dover))
——1913 The present state of the problem of gravitation *Phys. Z.* **14** 1249–66
——1916 The foundation of the General Theory of Relativity (in Einstein A, Lorentz H A, Weyl H and Minkowski H 1923 *The Principle of Relativity* (New York: Dover))
——1918 Principles of General Relativity *Ann. Phys., Lpz.* **55** 241–5
——1931 *Preuss. Akad. Wiss. Berlin Sitzungsber.* 235
——1954 *Relativity* (London: Methuen)
——1956 *The Meaning of Relativity* (London: Chapman and Hall)
——1969 Autobiographical notes in *Albert Einstein* ed P A Schilpp (New York: Open Court)
Einstein A and Grossmann M 1914 Draft of a generalised relativity theory and a theory of gravitation *Z. Math. Phys.* **62** 225–61
Einstein A and Infeld L 1961 *The Evolution of Physics* (New York: Simon and Schuster)
Einstein A, Lorentz H A, Weyl H and Minkowski H 1923 *The Principle of Relativity* (New York: Dover)
Ellis G F R 1978 Is the Universe expanding? *Gen. Relativ. Grav.* **9** 87–94
Ellis G F R and Schmidt B G 1977 *Gen. Relativ. Grav.* **8** 915
Eötvös R V 1889 Uber die Anziehung der Erde auf verschiedene Substanzen *Math. Naturwiss. Ber. Ung.* **8** 65–8
Eötvös R V, Pekar D and Fekete E 1922 Beiträge zum Gesetze der Proportionalität von Trägheit und Gravität *Ann. Phys., Lpz.* **68** 11–66
Feyerabend P 1970 Consolations for the specialist in *Criticism and the Growth of Knowledge* ed I Lakatos and A Musgrave (Cambridge: Cambridge University Press)

——1980 Zahar on Mach *Br. J. Phil. Sci.* **31** 273–82
——1981 *Philosophical Papers* (Cambridge: Cambridge University Press)
Feynman R 1965 *The Character of Physical Law* (Cambridge, MA: MIT Press)
Fine A 1986 Unnatural attitudes *Mind* **378** 149–79
Fock V A 1959 *The Theory of Space, Time and Gravitation* (Oxford: Pergamon)
van Fraassen B C 1970 *An Introduction to the Philosophy of Space and Time* (New York: Random House)
——1980 *The Scientific Image* (Oxford: Clarendon)
Freedman D Z and van Nieuwenhuizen P 1985 The hidden dimensions of spacetime *Scientific American* **252** 62–9
Friedman M 1973 Relativity principles and absolute objects in *Space, Time and Geometry* ed P Suppes (Dordrecht: Reidel) pp 296–320
——1983 *The Foundations of Spacetime Theories* (Princeton: Princeton University Press)
Friedmann A 1922 Uber die Krümmung des Raumes *Z. Phys.* **10** 377
Gardner M 1979 *The Ambidextrous Universe* (New York: Scribner)
Gardner M R 1977 Relationism and relativity *Br. J. Phil. Sci.* **28** 215–33
Geroch R P 1968 What is a singularity in general relativity? *Ann. Phys., NY* **48** 526–40
——1977 Prediction in general relativity in *Foundations of Spacetime Theories* ed J Earman, C Glymour and J Stachel (Minneapolis: University of Minnesota Press)
Geroch R P and Horowitz G T 1979 Global structure of spacetimes in *General Relativity* ed S W Hawking and W Israel (Cambridge: Cambridge University Press)
Gibbons G W, Hawking S W and Siklos S T C 1985 *The Very Early Universe* (Cambridge: Cambridge University Press)
Gödel K 1949 An example of a new type of cosmological solution of Einstein's field equations of gravitation *Rev. Mod. Phys.* **21** 447–50
Goldberg S 1984 *Understanding Relativity* (Oxford: Oxford University Press)
Goodman N 1983 *Fact, Fiction and Forecast* (Cambridge, MA: Harvard University Press)
Gott J R 1976 Will the Universe expand for ever? *Scientific American* **234** 62–79
Graves J C 1971 *The Conceptual Foundations of Contemporary Relativity Theory* (Cambridge, MA: MIT Press)
Gray J 1979 *Ideas of Space* (Oxford: Oxford University Press)
Gregory S A and Thompson L A 1982 Superclusters and voids in the distribution of galaxies *Scientific American* **246** 88–97
Gribbin J 1984 The bishop, the bucket, Newton and the Universe *New Scientist* (December) 12–15
Grunbaum A 1973 *Philosophical Problems of Space and Time* 2nd edn (Dordrecht: Reidel)
Haber H E and Kane G L 1986 Is nature supersymmetric? *Scientific American* **254** 42–50
Hacking I 1983 *Representing and Intervening* (Cambridge: Cambridge University Press)
Hanson N R 1958 *Patterns of Discovery* (Cambridge: Cambridge University Press)

Hart L and Davies R D 1982 Motion of the Local Group of galaxies and isotropy of the Universe *Nature* **297** 191–6
Hawking S W 1966 *Singularities and the Geometry of Spacetime* (unpublished, available from Department of Mathematics and Theoretical Physics, Silver Street, Cambridge, UK)
——1980 *Is the End in Sight for Theoretical Physics?* Inaugral lecture (Cambridge: Cambridge University Press) (also in Bosclough J 1984 *Beyond the Black Hole* (London: Fontana))
——1983 The cosmological constant: discussion in *The Constants of Physics* ed W H McCrea and M J Rees (Cambridge: Royal Society/Cambridge University Press) p. 93
Hawking S W and Ellis G F R 1973 *The Large Scale Structure of Space-time* (Cambridge: Cambridge University Press)
Hawking S W and Israel W (ed) 1979 *General Relativity* (Cambridge: Cambridge University Press)
Healey R A 1981 Statistical theories, quantum mechanics and the directedness of time in *Reduction, Time and Reality* ed R A Healey (Cambridge: Cambridge University Press)
Hempel C G 1965 *Aspects of Scientific Explanation* (New York: Free Press)
Henbest N and Couper H 1982 *The Restless Universe* (London: George Philip)
Hesse M B 1954 *Science and the Human Imagination* (SCM Press)
——1961 *Forces and Fields* (London: Nelson)
——1974 *The Structure of Scientific Inference* (London: Macmillan)
——1980a The hunt for scientific reason in *Proc. Conf. Phil. Sci. Assoc., 1980* vol. 2 ed P Asquith and R Giere (East Lansing, MI: Michigan State University Press)
——1980b *Revolutions and Reconstructions* (Bloomington, IN: Indiana University Press)
Hiebert E 1970 The genesis of Mach's early views on atomism in *Ernst Mach* ed R S Cohen and R J Seeger (Dordrecht: Reidel)
——1976a An appraisal of the work of Ernst Mach in *Motion and Time, Space and Matter* ed P K Machamer and R G Turnbull (Columbus, OH: Ohio State University Press)
——1976b Introduction in Mach E *Knowledge and Error* (Dordrecht: Reidel)
Hinckfuss I 1975 *The Existence of Space and Time* (Oxford: Oxford University Press)
Holton G 1970 Mach, Einstein and the search for reality in *Ernst Mach* ed R S Cohen and R J Seeger (Dordrecht: Reidel)
——1973 *Thematic Origins of Scientific Thought* (Cambridge, MA: Harvard University Press)
Hooker C A 1971 The relational doctrines of space and time *Br. J. Phil. Sci.* **22** 97–130
Horwich P 1975 On some alleged paradoxes of time travel *J. Phil.* **72** 432–44
——1978 On the existence of time, space and space-time *Nous* **12** 397
Hoyle F 1975 *Astronomy and Cosmology* (San Francisco: W H Freeman)
Hoyle F and Narlikar J 1974 *Action at a Distance in Physics and Cosmology* (San Franciso: W H Freeman)
Hubble E P 1936 The luminosity function of nebulae *Astrophys. J.* **84** 270–95

Hull D 1978 Planck's principle *Science* **202** 717
Jammer M 1969 *Concepts of Space* 2nd edn (Cambridge, MA: Harvard University Press)
——1974 *The Philosophy of Quantum Mechanics* (New York: Wiley)
Kerr R P 1963 Gravitational field of a spinning mass *Phys. Rev. Lett.* **11** 237
Kline M 1985 *Mathematics and the Search for Knowledge* (Oxford: Oxford University Press)
Kretschmann E 1917 On the physical meaning of the relativity postulate *Ann. Phys., Lpz.* **53** 574–614
Kuhn T S 1962 *The Structure of Scientific Revolutions* (Chicago: University of Chicago Press)
——1970 Reflections on my critics in *Criticism and the Growth of Knowledge* ed I Lakatos and A Musgrave (Cambridge: Cambridge University Press)
——1974 Second thoughts on paradigms in *The Structure of Scientific Theories* ed F Suppe (Urbana, IL: University of Illinois Press)
——1977 *The Essential Tension* (Chicago: University of Chicago Press)
Kursunoglu B 1962 *Modern Quantum Theory* (San Francisco: W H Freeman)
Lacey H M 1970 The scientific intelligibility of absolute space *Br. J. Phil. Sci.* **21** 317–42
Lakatos I 1970 Falsification and the methodology of scientific research programmes in *Criticism and the Growth of Knowledge* ed I Lakatos and A Musgrave (Cambridge: Cambridge University Press)
Lakatos I and Musgrave A (ed) 1970 *Criticism and the Growth of Knowledge* (Cambridge: Cambridge University Press)
Lanczos C 1974 *The Einstein Decade* (London: Elek Science)
Landau L D and Lifshitz E M 1975 *The Classical Theory of Fields* (Oxford: Pergamon)
Laudan L 1976 The methodological foundations of Mach's anti-atomism in *Motion and Time, Space and Matter* ed P K Machamer and R G Turnbull (Columbus, OH: Ohio State University Press)
——1986 *Science and Values* (Berkeley, CA: University of California Press)
Layzer D 1975 The arrow of time *Scientific American* **233** 56–69
Lewis D 1976 The paradoxes of time travel *Am. Phil. Q.* **13** 145–52
Lucas J R 1973 *A Treatise on Time and Space* (London: Methuen)
McCrea W H and Rees M J (ed) 1983 *The Constants of Physics* (Cambridge: Royal Society/Cambridge University Press)
Mach E 1911 *Conservation of Energy* (LaSalle, IL: Open Court) (first published 1872, Prague)
——1943 *Popular Scientific Lectures* 5th edn (LaSalle, IL: Open Court) (first published 1895)
——1959 *The Analysis of Sensations* (New York: Dover) (first published 1886, Jena)
——1960 *The Science of Mechanics* 6th edn (LaSalle, IL: Open Court) (first published 1883, Leipzig)
——1976 *Knowledge and Error* (Dordrecht: Reidel) (first published 1905, Leipzig)
Machamer P K and Turnbull R G (ed) 1976 *Motion and Time, Space and Matter* (Columbus, OH: Ohio State University Press)

Mackie J L 1980 In defence of induction in *Perception and Identity* ed G F MacDonald (London: Macmillan)
——1981 Space and time in *Space, Time and Causality* ed R Swinburne (Dordrecht: Reidel, 1984)
Malament D 1977 Observationally indistinguishable spacetimes in *Foundations of Spacetime Theories* ed J Earman, C Glymour and J Stachel (Minneapolis: University of Minnesota Press)
——1981 Newtonian gravity, limits, and the geometry of space in *Pittsburgh Studies in Philosophy of Science* ed R Colodny (Pittsburgh University Press, 1986)
Margenau H 1950 *The Nature of Physical Reality* (New York: McGraw-Hill)
Maxwell G 1962 The ontological status of theoretical entities in *Minnesota Studies in the Philosophy of Science* vol. III ed H Feigl and G Maxwell (Minneapolis: University of Minnesota Press)
Mehra J (ed) 1973 *The Physicist's Conception of Nature* (Dordrecht: Reidel)
Mellor D H 1981 *Real Time* (Cambridge: Cambridge University Press)
Misner C W, Thorne K S and Wheeler J A 1973 *Gravitation* (San Francisco: W H Freeman)
Nagel E 1961 *The Structure of Science* (London: Routledge and Kegan Paul)
Narlikar J 1978 *The Structure of the Universe* (Oxford: Oxford University Press)
Nerlich G 1973 *The Shape of Space* (Cambridge: Cambridge University Press)
——1979 What can geometry explain? *Br. J. Phil. Sci.* **30** 69
Newburgh R G 1973 Comments on Mach's principle *Br. J. Phil. Sci.* **24** 263
Newton I 1729 *Principles of Natural Philosophy* (translation by A Motte) (Dawsons)
——1934 *Principles of Natural Philosophy* (translation by F Cajori) (Berkeley, CA: University of California Press)
Newton-Smith W H 1978 The underdetermination of theory by data *Proc. Aristotelean Soc.* suppl. vol. **I–II** 71–91
——1980 *The Structure of Time* (London: Routledge and Kegan Paul)
——1981 *The Rationality of Science* (London: Routledge and Kegan Paul)
North J D 1965 *The Measure of the Universe* (Oxford: Clarendon)
O'Hanlon J and Tupper B 1972 The Brans–Dicke theory of gravitation *Nuovo Cimento* B **7** 305
Oszvàth I and Schücking E 1969 The finite rotating universe *Ann. Phys., NY* **55** 166–204
Pais A 1982 *Subtle is the Lord* (Oxford: Oxford University Press)
——1986 *Inward Bound* (Oxford: Oxford University Press)
Papineau D 1979 *Theory and Meaning* (Oxford: Oxford University Press)
Pauli W 1958 *Theory of Relativity* (London: Pergamon)
Penrose R 1974 Twistors and particles: an outline in *Quantum Theory and the Structures of Time and Space* ed L Castell, M Drieschner and C F Weizsäcker (Munich: Carl Hanser)
——1979 Singularities and time-asymmetry in *General Relativity* ed S W Hawking and W Israel (Cambridge: Cambridge University Press)
Planck H 1949 *Scientific Autobiography* (New York: Philosophical Library)
Popper K R 1959 *The Logic of Scientific Discovery* (London: Hutchinson)
——1963 *Conjectures and Refutations* (London: Routledge and Kegan Paul)

——1970 Normal science and its dangers in *Criticism and the Growth of Knowledge* ed I Lakatos and A Musgrave (Cambridge: Cambridge University Press)

Putnam H 1975 *Philosophical Papers* (Cambridge: Cambridge University Press)

Pyenson A 1985 *The Young Einstein* (Bristol: Adam Hilger)

Quine W V O 1953 *From a Logical Point of View* (Cambridge, MA: Harvard University Press)

Raine D J 1975 Mach's principle in General Relativity *Mon. Not. R. Astron. Soc.* **171** 507–28

——1981 *The Isotropic Universe* (Bristol: Adam Hilger)

Raychaudhuri A K 1979 *Theoretical Cosmology* (Oxford: Clarendon)

Reasenberg R D 1983 Constancy of *G* and other gravitational experiments in *The Constants of Physics* ed W H McCrea and M J Rees (Cambridge: Royal Society/Cambridge University Press) p. 17

Redhead M 1981 Causality, relativity and E.P.R. paradox in *Space, Time and Causality* ed R Swinburne (Dordrecht: Reidel, 1984)

Reichenbach H 1942 *From Copernicus to Einstein* (New York: Dover)

——1958 *The Philosophy of Space and Time* (New York: Dover)

——1969 The philosophical significance of the theory of relativity in *Albert Einstein* ed P A Schilpp (LaSalle, IL: Open Court)

——1971 *The Direction of Time* (Berkeley, CA: University of California Press)

Reindhardt M 1973 Mach's principle—a critical review *Z. Naturforsch.* **28A** 529–37

Ridley B K 1976 *Time, Space and Things* (Harmondsworth: Pelican)

Rindler W 1977 *Essential Relativity* 2nd edn (Berlin: Springer)

Robinson F H 1973 *Macroscopic Electromagnetism* (Oxford: Pergamon)

Russell B 1903 *The Principles of Mathematics* (London: Allen and Unwin)

——1927 *The Analysis of Matter* (London: Kegan Paul)

Sachs M 1972 On the Mach principle *Br. J. Phil. Sci.* **23** 117

——1976 On the logical status of equivalence principles *Br. J. Phil. Sci.* **27** 225

Salmon W C 1975 *Space, Time and Motion* (Berkeley, CA: University of California Press)

Schilpp P A (ed) 1969 *Albert Einstein* (LaSalle, IL: Open Court)

Sciama D W 1953 On the origin of inertia *Mon. Not. R. Astron. Soc.* **113** 34

——1961 *The Unity of the Universe* (New York: Doubleday)

——1969 *The Physical Foundations of General Relativity* (New York: Doubleday) and (London: Heinemann, 1972)

——1971 *Modern Cosmology* (Cambridge: Cambridge University Press)

——1973 The Universe as a whole in *The Physicist's Conception of Nature* ed J Mehra (Dordrecht: Reidel)

Sciama D W, Waylen P C and Gilman R C 1969 Generally covariant integral formulation of Einstein's field equations *Phys. Rev.* **187** 1762

Sellars W 1963 *Science, Perception and Reality* (London: Routledge and Kegan Paul)

Shapiro I I 1964 Fourth test of general relativity *Phys. Rev. Lett.* **13** 789

Sklar L 1974 *Space, Time and Spacetime* (Berkeley, CA: University of California Press)

Stachel J 1972 The rise and fall of geometrodynamics in *Proc. Conf. Phil. Sci. Assoc., 1972* ed R S Cohen (Dordrecht: Reidel)
Stein H 1967 Newtonian spacetime *Texas Quarterly* **10** 174–200
——1968 On Einstein–Minkowski space–time *J. Phil.* **65** 5–22
——1977 Some philosophical prehistory of general relativity in *Foundations of Spacetime Theories* ed J Earman, C Glymour and J Stachel (Minneapolis: University of Minnesota Press)
Suppes P 1957 *Introduction to Logic* (New York: Van Nostrand)
Swinburne R 1968 *Space and Time* (London: Macmillan)
—— (ed) 1984 *Space, Time and Causality* (Dordrecht: Reidel)
Synge J L 1966 *Relativity: the General Theory* (Oxford: Oxford University Press)
Teplitz C and Teplitz V L 1983 The future of the Universe *Scientific American* **248** 74–85
Tipler F J 1978 Mach's principle in spatially homogeneous spacetimes *Phys. Lett.* **68A** 313
Tomozawa T 1972 Mach's principle *Found. Phys.* **2** 27
Trautman A 1967 Foundations and current problems of general relativity in *Lectures on General Relativity* ed A Trautman, F Pirani and H Bondi (Englewood Cliffs, NJ: Prentice-Hall)
Truesdell C 1984 *The Elements of Continuum Mechanics* (Berlin: Springer)
Wald R M 1984 *General Relativity* (Chicago: University of Chicago Press)
Weinberg S 1972 *Gravitation and Cosmology* (New York; Wiley)
——1977 *The First Three Minutes* (New York: Bantam)
Weingard R 1977 On cracking that nut, absolute space *Phil. Sci.* **44** 288
——1979 Some philosphical aspects of black holes *Synthese* **42** 191
Weyl H 1922 *Space, Time and Matter* (London: Methuen)
——1952 *Symmetry* (Princeton, NJ: Princeton University Press)
Wheeler J A 1964 Mach's principle as a boundary condition for Einstein's equations in *Gravitation and Relativity* ed H Y Chiu and W F Hoffman (New York: Benjamin)
Whewell W 1851 Of the transformation of hypotheses *Proc. Camb. Phil. Soc.* **9** 139
Whitrow G J 1980 *The Natural Philosophy of Time* (Oxford: Oxford University Press)
Will C M 1979 The confrontation between gravitation theory and experiment in *General Relativity* ed S W Hawking and W Israel (Cambridge: Cambridge University Press)
——1981 *Theory and Experiments in Gravitational Physics* (Cambridge: Cambridge University Press)
Wittgenstein L 1953 *Philosophical Investigations* (Oxford: Blackwell)
Woodward J F and Yourgrau W 1972a The incompatibility of Mach's principle and the principle of equivalence *Br. J. Phil. Sci.* **23** 111
——1972b Mach's principle *Br. J. Phil. Sci.* **24** 264
——1975 Mach's principle: micro- or macrophysical? *Br. J. Phil. Sci.* **27** 137
Yodzis P, Seifert H and Müller zum Hagen H 1973 On the occurrence of naked singularities in general relativity I *Comm. Math. Phys.* **34** 135–48

——1974 On the occurrence of naked singularities in general relativity II *Comm. Math. Phys.* **37** 29–40

Zahar E 1977 Mach, Einstein and the rise of modern science *Br. J. Phil. Sci.* **28** 195–213

——1980 Einstein, Meyerson and the role of mathematics in physical discovery *Br. J. Phil. Sci.* **31** 1–43

——1981 Second thoughts about Machian positivism *Br. J. Phil. Sci.* **32** 267–76

Zaret D 1979 Absolute space and conventionalism *Br. J. Phil. Sci.* **30** 180–97

Zel'dovich Y B 1972 A hypothesis unifying the structure and the entropy of the universe *Mon. Not. R. Astron. Soc.* **160** 1P–3P

——1979 Cosmology and the early universe in *General Relativity* ed S W Hawking and W Israel (Cambridge: Cambridge University Press)

Index

QUEEN MARY
COLLEGE
LIBRARY

WITHDRAWN
FROM STOCK
QMUL LIBRARY

D1345726